Focus on Teens in Trouble

A Reference Handbook

TEENAGE PERSPECTIVES

Focus on Teens in Trouble

A Reference Handbook

Daryl Sander

Professor of Counseling and Educational Psychology University of Colorado–Boulder

Santa Barbara, California
Oxford, England

Library of Congress Cataloging-in-Publication Data
Sander, Daryl.
Focus on teens in trouble : a reference handbook / Daryl Sander.
p. cm.—(Teenage perspectives)
Includes bibliographical references and index.
Summary: Discusses such aspects of juvenile delinquency as gangs, drug abuse, running away, and crime. Provides resources for additional information.
1. Juvenile deliquency—United States—Juvenile literature.
2. Juvenile deliquents—United States—Juvenile literature.
3. Juvenile justice, Administration of—United States—Juvenile literature. [1. Juvenile delinquency.] I. Title. II. Series.
HV9104.S315 1991 364.3'6'0973—dc20 90-24777

ISBN 0-87436-207-5 (alk. paper)

98 97 96 95 94 93 92 91 10 9 8 7 6 5 4 3 2 1

ABC-CLIO, Inc.
130 Cremona Drive, P.O. Box 1911
Santa Barbara, California 93116-1911

Clio Press Ltd.
55 St. Thomas' Street
Oxford, OX1 1JG, England

This book is printed on acid-free paper ♾.
Manufactured in the United States of America

This work is dedicated
to Mary and our son, William.

Contents

Foreword

An old Chinese curse goes, "May you live in exciting times." Today's young people have grown up under such a curse—or blessing. They live in a world that is undergoing dramatic changes on every level—social, political, scientific, environmental, technological. At the same time, while still in school, they are dealing with serious issues, making choices and confronting dilemmas that previous generations never dreamed of.

Technology, especially telecommunications and computers, has made it possible for young people to know a great deal about their world and what goes on in it, at least on a surface level. They have access to incredible amounts of information, yet much of that information seems irrelevant to their daily lives. When it comes to grappling with the issues that actually touch them, they may have a tough time finding out what they need to know.

The Teenage Perspectives series is designed to give young people access to information on the topics that are closest to their lives or that deeply concern them—topics like families, school, health, sexuality, and drug abuse. Having knowledge about these issues can make it easier to understand and cope with them, and to make appropriate and beneficial choices. The books can be used as tools for researching school assignments, or for finding out about topics of personal concern. Adults who are working with young people, such as teachers, counselors, librarians, and parents, will also find these books useful. Many of the references cited can be used for planning information or discussion sessions with adults as well as young people.

Ruth K. J. Cline
Series Editor

Preface

Anyone who reads this book already knows that we are faced with a serious crime problem in this country. What may be less well known, however, is the extent to which youths under the age of 25, or even in their teens, are represented in the statistics that appear in the daily news media. By 1982, the leading cause of death among 15- to 24-year-olds was no longer auto accidents, but homicide. The greatest increase in the homicide rate was among 15- to 19-year-old black males (Nancy Needham, "Kids Who Kill and Are Killed," *NEA Today* 6, no. 1 [February 1988]: 10–11). We read estimates of the economic costs of such criminal behavior but seldom explore fully the social and psychological costs, to the young people involved and to the families as well.

The seriousness of youth crime is underscored by the current public outrage, demands for stricter enforcement of laws, and demands for harsher penalties for young persons convicted of crime. Though historically ours has been a society that has treated juvenile crime differently from adult crime, the times are changing. A recent change in Indiana law requires that offenders as young as 16 be tried as adults when charged with either the sale of or possession of sawed-off shotguns. Illinois recently changed its law so that 15- and 16-year-olds are tried as adults when charged with sexual assault, rape, armed robbery, or murder. Kids' crime is still crime, and the public has said, "Get tough!"

This book provides detailed information about youth offenses of all kinds, an examination of the full scope and depth of the problem of young people who run afoul of the law, and insights into the complex factors that lead some youngsters in this self-defeating direction. Special attention is given to the development of adolescents and their culture. Youth crimes discussed in this book range from the less serious, such as vandalism in the public park, to the bloodshed and violence of gang wars and drug trafficking.

The resources section for each chapter presents a collection of fiction, nonfiction (both books and articles), and nonprint materials related to that chapter's topic. (Note that nonprint materials are often available on a rental basis through college and university film libraries. Because of space limitations, only one film library has been listed for each film or video, though indeed the materials are often available for rental from a number of university media centers. Readers are therefore encouraged to check with a college or university film library near them concerning the availability of films and videos for rental.) A listing of relevant organizations and, in some cases, hotline numbers completes each chapter.

The purpose of this book is not to provide a pat formula for why teens get into trouble but rather to make readers question why this problem has reached the magnitude it has today. In the belief that new solutions to this problem can be found only in knowledge and creative ideas, this book provides readers with a wide range of sources of information.

CHAPTER 1

Adolescents

My father said softly, "Sweetheart, you'd feel better if you were doing something all day instead of hanging around."

The screams came back full force. "What do you want me to do?" she shrieked, her voice high and bitter and raging. "Sell hamburgers at Burger King? Type invoices at Bloomingdales?" She was gesturing with the bread knife now.

My mother drew back from the table mesmerized by the swings of that knife. Her fingers knotted around each other like crochet of the flesh.

Ashley spat out, "I am not common. I will not do common things!"

Caroline B. Cooney, *Don't Blame the Music* (New York: Pacer Books, 1987), 114–115.

All teenagers are defining themselves as they move into young adulthood. Their self-concept changes to accommodate social development and the physical changes triggered by puberty. Some experience anger and rebellion in regard to parents or other authority figures. All are seeking to become more independent of family and free from parental supervision (Onyehalu, 7–10). For most, this transition period—adolescence—is a happy and trouble-free period of life. For a few, the teen years are troublesome, sometimes marked by antisocial or illegal behavior.

Although this book is largely about teens in trouble, it should be emphasized that the majority of teenagers are happy, healthy young people who are relatively trouble-free. Dr. Daniel Offer recently reported that in his research with high school students only 20 percent or so were troubled or rebelling against parents and the adult world. He said that "the vast majority of teenagers function well, enjoy good relationships with their families and friends, and accept the values of the larger society" (*New York Times*, 3 February 1987: C-7).

Statistics about Teens in Trouble

- Of all criminal arrests in 1988, 21 percent were people 18 years and under (FBI, 178)
- Of all felonies committed in 1980, 40 percent were committed by juveniles (*Study Guide*, 2)
- Eighty percent of all juveniles who were held in detention in 1987 were males and 20 percent were females (*Sourcebook*, 595)

Definitions

Self-concept. "One's own definition of who one is, including attributes, emotions, abilities, faults, etc." (Corsini, 595).

Adolescence. The period of rapid growth—both psychological and biological—between childhood and adulthood (Atwater, 1).

Juvenile. A legal term referring to any underage person, usually under 21 years; a minor.

Delinquency. "Failure, omission, or violations of a law or duty" (Black, 385).

Factors Causing Delinquency

The causes of delinquent or antisocial behavior among adolescents have not been convincingly established. However, experts have offered theories involving causal factors that range from personality problems to family, social, and economic considerations. There is some research evidence supporting each of these as a potential cause of delinquency (Reed, 216–217). Among those factors most frequently mentioned are: (1) low self-esteem of the teenager, (2) changing family structure, (3) social changes, (4) violence shown in TV and movies, (5) school failure, and (6) poverty. Delinquent behavior can probably be best attributed to differing combinations of these factors in each delinquent's life (Atwater, 283–284).

LOW SELF-ESTEEM

Self-concept and self-esteem have long been important parts of the literature on personality development (Rogers, 215–219). Self-concept is how persons see themselves; self-esteem the value placed upon this self. Low self-esteem results when adolescents see themselves in largely negative terms—incapable, unappreciated, or unloved. A leading psychologist says that "delinquent youths tend to show lower self-esteem than do non-delinquent youths" (Rice, 219). Another expert claims that "feelings of low self-esteem are often acted out in self-destructive, delinquent behavior that involved violence" (Yablonsky and Haskell, 367). According to these views, youths with low self-esteem behave in ways that match this evaluation, and the result is behavior that can best be described as self-punishing or self-destructive. It is not totally clear which is cause and which is effect.

CHANGING FAMILY STRUCTURE

Today's teenagers are likely to be growing up in a family home that is quite different from the traditional pattern of several decades ago. In the past the usual family consisted of two parents (usually the biological parents) and children living in a home where the mother was full-time homemaker and caregiver. Parental supervision of the children was direct and constant. Today, there are an estimated seven million single-parent homes (Kobliner, 63). Many of today's teens have been "latch-key" children who returned home from school to an environment lacking parental supervision for several hours daily until a working parent appeared. Stepfamily arrangements are much more common, and these are sometimes a source of tension or conflict.

It is estimated that more than half of the women today who have children under three are working outside their homes. This trend appears to be increasing from one decade to the next. In 1970 about 43 percent of women with school-age children were working outside the home, but this had increased to 51 percent by 1980 (U.S. Department of Commerce, 6–9). Clearly this indicates that there is less time for parental supervision and that there are fewer opportunities for positive parent role modeling than in previous decades.

SOCIAL CHANGES

In addition to the decline in the proportion of traditional nuclear families, today's teens may live in some degree of social isolation. Frequent moves to accommodate a parent's change in jobs sometimes leave the family in semi-isolation in a new residence. Relatives and old friends who might provide support and stability in a family may live thousands of miles away. Some research indicates a connection between psychological stress and mobility, and isolation of families (Conger and Petersen, 199). Teenagers find themselves in a much different home and social environment from what their parents experienced. "When the developmental experiences that shape our personalities and the social changes that must be confronted vary markedly from adults to young people, from parents to their children, generational differences in cultural values and outlook—even in knowledge—tend to be magnified" (ibid., 200).

TELEVISION, MOVIES, AND VIOLENCE

Although many have expressed opinions about television's influence on behavior, the research evidence is not entirely consistent. It is a fact that children as young as three or four are averaging four hours of TV viewing per day (Singer, 815). In general, the amount of time that children spend watching TV increases throughout the elementary school years. One researcher concludes that "children spend more time in this country watching television than they ever spend in school" (ibid.). It is also quite clear that violence and aggression are depicted in a large majority of American TV programs (Liebert and Sprafkin, 117). Likewise, violence as a means of solving problems is featured in many R-rated films, which are easily viewed by teens and children in neighborhood theaters (Wloszczyna, 4D).

The direct connection between watching TV violence and subsequent violent behavior by the viewer is somewhat less clear. However, after reviewing several large studies of this connection, researchers concluded that "preference for violent programs was associated with aggressive and violent acts, petty delinquency, defiance of parents, political protest, and serious delinquency, defined as involving conflict with the law" (Apter and Goldstein, 186). Although some behavioral scientists do not agree with that conclusion, an increasing amount of evidence points to a connection (Liebert and Sprafkin, 135–161). Further, recent research suggests that teens often don't stop to consider alternatives to violence when dealing with interpersonal conflict (Landers, 36–37).

Another scholar has taken the view that not only does today's television promote violent behavior, but also that its wider effect on values and attitudes has been largely negative (Silber, 64–68). He states that television "teaches us that violence is normal and enjoyable . . . that happiness is no more than instant gratification; that sexual intercourse is to be engaged in by casual strangers or even enjoyed as a spectator sport" (ibid., 69). Thus there appears to be a case for regarding television's influence on children and teens as being both significant and largely negative.

SCHOOL FAILURE

Every state has mandatory school attendance laws, usually requiring youths to attend school until age 16 or graduation. Further, most families view success at school as a necessary step to responsible adulthood. Yet for some teens, school is boring and unpleasant, and becomes a series of failures. In New York, dropout rates for minority youth over 14 years of age range from 30 percent to 62 percent, depending upon whose data are used (Berger, B1). School failures and school dropout have been viewed by many as closely linked with antisocial behavior. "The relationship between school failure and criminal behavior is a recurrent theme in theories of delinquency. Eventual dropouts have been found to have considerably higher rates of delinquency during high school than the graduates . . . " (Thornberry, Moore, and Christensen, 56). Such relationships should be viewed with caution, however, since it is seldom clear which factor is the cause and which the effect.

POVERTY

The connection between poverty and crime has been studied by many experts. Although there appears to be some relationship for some crimes, the connection is very complex (Elliott, Huizinga, and Menard, 193; Reid, 153–158; Yablonsky and Haskell, 375–383). For example, it is well known that certain types of youth crimes occur at a higher rate in the urban ghetto than in rural New England. However, there are many other factors that might explain these differential rates, e.g., police presence and arrest policies, types of schools and other neighborhood organizations, and sense of community among the residents. Poverty is oppressive; families are often unable to better themselves despite their best efforts. Teenagers who are reared in poverty sometimes find themselves trapped in an intergenerational poverty cycle (Kantrowitz, 78). The socioeconomic level of families is but one factor in the complex social structure in which youth crime occurs (Sanders, 41–56). Yet, poverty as a contributing factor to teen crime can hardly be ignored.

Summary

The vast majority of teenagers do not experience trouble with the law. Yet some do, and their plight is one of concern to society. The causes of juvenile crime are many and varied, and experts in the fields of behavioral science and criminology believe that combinations of factors probably contribute to the determination of which youth are likely to get into trouble and which are not.

REFERENCES

Apter, Steven J., and Arthur Goldstein, eds. *Youth Violence: Program and Prospects*. New York: Pergamon Press, 1986.

Atwater, Eastwood. *Adolescence*. Englewood Cliffs, NJ: Prentice-Hall, 1983.

Bender, Arnold, Gilbert Geis, and Dickson Bruce. *Juvenile Delinquency—Historical, Cultural, and Legal Perspectives*. New York: Macmillan, 1988.

Berger, Joseph. "Dropout Plans Not Working, Study Finds," *New York Times*, 16 May 1990: B-1.

Black, Henry C. *Black's Law Dictionary*. 5th ed. St. Paul: West Publishing, 1979.

Brazelton, T. Berry. "Working Parents," *Newsweek* 113, no. 7 (13 February 1989): 66–70.

Conger, John J., and Anne Petersen. *Adolescence and Youth: Psychological Development in a Changing World*. 3d ed. New York: Harper & Row, 1984.

Cooney, Caroline B. *Don't Blame the Music*. New York: Pacer Books, 1987.

Corsini, Raymond J., and Danny Wedding. *Current Psychotherapies*. 4th ed. Itasca, IL: Peacock Publishers, 1989.

Elliott, Delbert S., David Huizinga, and Scott Menard. *Multiple Problem Youth*. New York: Springer-Verlag, 1989.

Federal Bureau of Investigation. *Uniform Crime Reports of the United States*. Washington, DC: U.S. Department of Justice, 1988.

Kantrowitz, Susan. "Breaking the Poverty Cycle," *Newsweek* 114, no. 20 (28 May 1990): 78.

Kobliner, Beth. "Solo Parenting Breeds a Keen Eye For Priorities," *Money* 18, no. 3 (March 1989): 63–64.

Landers, Susan. "Teens Rarely Consider Alternatives to Violence," *APA Monitor* 20, no. 10 (October 1989): 36–37.

Liebert, Robert M., and Joyce Sprafkin. *The Early Window: Effects of Television on Children and Youth*. 3d ed. New York: Pergamon Press, 1988.

Offer, Daniel. "The Mystery of Adolescence," *Adolescent Psychiatry* 14 (1987): 7–27.

Onyehalu, Anthony S. "Identity Crisis in Adolescence," *Adolescence* 16, no. 63 (Fall 1981): 629–632.

Reid, Sue T. *Crime and Criminology*. 5th ed. New York: Holt, Rinehart & Winston, 1988.

Rice, Phillip. *The Adolescent—Development, Relationships, and Cultures*. 4th ed. Boston: Allyn & Bacon, 1984.

Rogers, Dorothy. *Life-Span Human Development*. Monterey, CA: Brooks/Cole, 1982.

Sanders, William B. *Juvenile Delinquency*. New York: Holt, Rinehart & Winston, 1981.

"Science Watch," *New York Times*, 3 February 1987: C-7.

Silber, John. *Straight Shooting*. New York: Harper & Row, 1989.

Singer, Dorothy G. "A Time To Reexamine the Role of Television in Our Lives," *American Psychologist* 38, no. 7 (July 1983): 815–816.

Sourcebook of Criminal Justice Statistics—1988. Washington, DC: U.S. Department of Justice, 1989.

Study Guide, Crime File—Juvenile Offenders. Washington, DC: U.S. Department of Justice, 1985.

Thornberry, Terrance P., Melanie Moore, and R. L. Christensen. "The Effect of Dropping Out of High School on Subsequent Criminal Behavior," *Criminology* 23, no. 1 (February 1985): 3–18.

U.S. Department of Commerce. Bureau of the Census. *Population Profile of the United States*. Washington, DC: U.S. Government Printing Office, 1981. (Current Population Reports. Series P-20, No. 363.)

Wloszczyna, Susan. "R-Rated Films Are Often beyond Parental Discretion," *USA Today*, 12 June 1990: 4D.

Yablonsky, Lewis, and Martin Haskell. *Juvenile Delinquency*. 4th ed. New York: Harper & Row, 1988.

Resources
for Finding Out about Adolescents

Fiction

Belair, Richard. **Double Take.** New York: Morrow, 1979. 190p.

Robert believes that people are judged solely on personal appearance, which he believes is wrong, and so makes no effort to attract friends by keeping a neat appearance. He becomes interested in photography, and it is this hobby that teaches him there is beauty in all, and that people can be interested in the real you.

Bridgers, Sue Ellen. **Notes for Another Life.** New York: Bantam, 1982. 201p.

Wren and Kevin live with their grandparents because their father is in a mental institution and their mother prefers to have a career life of her own. They feel as if they must never ask anything from life because of the strangeness of their upbringing, but they learn that life and people can be worth trusting.

Cooney, Caroline. **Don't Blame the Music.** Berkeley, CA: Pacer Books, 1987. 172p.

Susan's sister Ashley returns home after failing in her attempt to be a rock star. Although Susan and her family try to help Ashley adjust to family life again, they finally accept the fact that Ashley needs more than their help for her bitterness and alienation.

Cooney, Linda A. **Sweet Sixteen.** New York: Scholastic, 1989. 219p.

A look at the developmental issues faced by three teenage girls. The social development of these teenagers is the focus of this book.

Crutcher, Chris. **Stotan!** New York: Greenwillow, 1986. 183p.

A group of four friends, athletes, complete their coach's week of endurance tests. They learn a lot about athletics, competing, each other, and going the whole distance.

DeClements, Barthe. **I Never Asked You To Understand Me.** New York: Viking, 1986. 138p.

Didi enters an alternative school after nearly flunking out of her old school. She becomes involved in parties and drugs, and makes friends with Stacy, whose home life is abusive. Didi's mother dies, and she goes to live with her grandmother, but she has learned some lessons about people and friendship that make her stronger.

Forshay-Lunsford, Cin. **Walk through Cold Fire.** New York: Delacorte, 1985. 224p.

Desiree is forced to spend the summer with her aunt when her father remarries, and she is attracted to the "outcast" teenagers in town, especially Billy. She identifies with the pushed-out feelings of the kids in the Outlaws group, and owes them her new sense of self-worth.

Greenberg, Jan. **The Pig-Out Blues.** New York: Farrar, Straus & Giroux, 1982. 121p.

Jodie is pushed by her mother to lose weight, but it is a part in *Romeo and Juliet* that is Jodie's incentive. Jodie is struggling to separate herself from her mother, and to make her own friends and values.

Klein, Norma. **Learning How To Fall.** New York: Bantam, 1989. 183p.

Seventeen-year-old Dustin tries to work out the uncertainties of identity and values in a confusing world. Becoming sexually involved contributes to his confusion and eventual emotional breakdown.

Major, Kevin. **Dear Bruce Springsteen.** New York: Delacorte, 1987. 134p.

A boy shares his problems, thoughts, triumphs, and feelings about adolescence and his life through a series of letters to his idol, Bruce Springsteen.

Malmgren, Dallin. **The Whole Nine Yards.** New York: Delacorte, 1986. 137p.

Storm spends the majority of his high school years goofing off with his friends and chasing after girls. He finally finds the one he really likes, Paula, but because of his immaturity, he loses her. The death of his best friend makes him see things in a new light.

Morgenroth, Barbara. **Tramps Like Us.** New York: Atheneum, 1979. 144p.

Intelligent Vanessa has not felt challenged by anything in her life. She meets Daryl, who encourages her to question her life and values, but her parents reject Daryl as a tramp. They run away together, and after being returned by the police, Vanessa feels confident she can confront the world on her terms.

———. **Will the Real Renie Lake Please Stand Up?** New York: Atheneum, 1981. 164p.

Renie goes to live with her father and his new family after getting into trouble while living with her mother and her mother's boyfriend. Renie feels she must keep adjusting her personality to get along in new situations, but finally she feels that she can be herself after making friends with a deaf boy.

Peck, Richard. **Princess Ashley.** New York: Delacorte, 1987. 205p.

Chelsea is in a new school and wants to be a part of things. She can't believe it when popular Ashley wants to be her friend, but Chelsea learns that friendship can be deceiving.

Pevsner, Stella. **Cute Is a Four-Letter Word.** New York: Houghton Mifflin, Clarion Books, 1980. 190p.

Eighth-grader Clara discovers that being cute isn't everything. She is surprised when she makes the pom-pom squad and is dated by the captain of the football team, but more is going on than Clara realizes.

Sieruta, Peter D. **Heartbeats and Other Stories.** New York: Harper & Row, 1989. 214p.

This collection of nine short stories describes many of the problems faced by teenagers today. The stories cover a wide range of teen emotional issues, including family conflict and identity.

Voigt, Cynthia. **The Runner.** New York: Atheneum, 1985. 181p.

Bullet comes from a family in which the father is stern and bitter, the mother noncommunicative, and the other two siblings have already left. He doesn't care about other people very much; running, but not competing, has become his life. He does learn to care some about others as he recognizes his racism, but there is still a sense of incompleteness in his life.

Nonfiction

Arnold, L. Eugene, ed. **Preventing Adolescent Alienation.** Lexington, MA: Lexington Books, 1983. 146p.

Alienation can't be eliminated among today's teenagers, but the authors represented here have many ideas about minimizing its effects. One of the writers suggests ways to help adolescents gain a sense of belonging and a feeling of being an important part of what goes on in their world.

Atwater, Eastwood. **Adolescence.** Englewood Cliffs, NJ: Prentice-Hall, 1983. 471p.

A very detailed account of many aspects of adolescence. Chapter 12, "Moral Development," and Chapter 13, "Juvenile Delinquency," are especially relevant to those who want a deeper understanding of teenagers who find themselves in trouble with the law.

Berger, Gilda. **Violence and the Media.** New York: Franklin Watts, 1989. 176p.

A comprehensive study of the state of violence in all media. She emphasizes the fact that we are so highly exposed to violence that we are no longer aware of it.

Bloomfield, Harold H. **Making Peace with Your Parents.** New York: Random House, 1983. 220p.

This book contains ideas, techniques, and exercises for improving teens' understanding of their parents. Psychiatrist Bloomfield presents some excellent ideas for improving communications with parents. This is especially evident in Chapter 3, titled "Expressing Anger and Love with Your Family."

Comfort, Alex, and Jane Comfort. **The Facts of Love—Living, Loving and Growing Up.** New York: Ballantine, 1980. 157p.

This book on human sexuality was written especially for teenagers. It provides information on basic physical development as well as the emotional and social aspects of adolescent sexuality, peer pressure, and parents' likely reactions.

Conger, John J., and Anne Petersen. **Adolescence and Youth: Psychological Development in a Changing World.** 3d ed. New York: Harper & Row, 1984. 732p.

This book contains comprehensive references and research citations on adolescence. The chapter on "Alienation and Delinquency" is especially relevant to readers of *Focus on Teens in Trouble.*

Connolly, Patrick. **Love, Dad.** Kansas City, MO: Andrews, McMeel, and Parker, 1985. 200p.

Connolly wrote this collection of letters to his sons during their growing-up years. They incorporate cartoons and caricatures with fatherly wisdom and love.

Coombs, H. Samm. **Teenage Survival Manual: How To Reach "20" in One Piece (and Enjoy Every Step of the Journey).** 4th ed. Lagunitas, CA: Discovery Books, 1989. 235p.

This advice manual for teenagers covers a wide variety of the daily traumas that teens face. The emphasis is on personal integrity and the courage to keep trying.

Elkind, David. **All Grown Up & No Place To Go.** Reading, MA: Addison-Wesley, 1984. 232p.

This book is helpful in understanding the frustrations and inner conflicts of teens. Chapter 3, "Perils of Puberty," is an excellent but brief treatment of sexual development and related social problems.

Frey, Diane, and C. Jesse Carlock. **Enhancing Self Esteem.** Muncie, IN: Accelerated Development, 1984. 317p.

Developing teenagers who want to build self-esteem can follow the systematic approach described here. The authors emphasize understanding the self and viewing one's self more positively.

Kelly, Jeffrey. **Social Skills Training: A Practical Guide for Interventions.** New York: Springer, 1982. 272p.

A practical guide to learning new skills of communication.

Keniston, Kenneth. **The Uncommitted: Alienated Youth in American Society.** New York: Harcourt, 1965. 500p.

Psychologist Keniston explains *how* and *why* young people in the 1960s became so alienated. He describes in detail the social conditions that foster alienation.

Keniston, Kenneth. **Youth and Dissent: The Rise of a New Opposition.** New York: Harcourt, 1971. 403p.

Keniston's analysis of alienation among young people was extended and updated in this volume. He pays particular attention to economic and social variables that interact with personality dynamics to produce anger and frustration among youth.

Klein, David, and Marymae E. Klein. **Your Parents and Your Self.** New York: Scribner's, 1986. 166p.

An account of how all people (especially the teens to whom this is addressed) are products of genetic endowment. The first chapter is aptly titled "What Can You Blame on Your Parents?", and subsequent chapters deal with sociability, personality, career interests, and the like.

Kolodny, Nancy, Robert Kolodny, and Thomas Bratter. **Smart Choices.** Boston: Little, Brown, 1986. 176p.

The teen years involve making choices and important decisions on a variety of topics from drugs and sex to social and career concerns. This book deals with numerous developmental and social dilemmas and decision points. It emphasizes practical approaches to decision making.

Kolodny, Robert C., Nancy J. Kolodny, Thomas E. Bratter, and Cheryl A. Deep. **How To Survive Your Adolescent's Adolescence.** Boston: Little, Brown, 1984. 350p.

Information about the types of stress faced by teenagers and how they attempt to cope.

Levine, Mel. **Keeping Ahead in School: A Student's Book about Learning Abilities and Learning Disorders.** Cambridge, MA: Educators Publishing Service, 1990. 320p.

Aimed specifically at teens, this book provides basic information about individual differences among learners. The author helps teens to understand what learning disabilities are, and how they are different from stupidity or low intelligence.

Seebald, Hans. **Adolescence: A Social Psychological Analysis.** 3d ed. Englewood Cliffs, NJ: Prentice-Hall, 1984. 352p.

Extensive information about the development of adolescents. Two chapters deserve special mention—Chapter 8, "The Role of Television and Other Mass Media," and Chapter 10, "Inside the Teen Subculture."

Snider, Dee, and Philip Bashe. **Teenage Survival Guide.** Garden City, NY: Doubleday, 1984. 240p.

Dee Snider has written directly to teens in this book, and he tells them what they need to hear, in their own language. The chapter entitled "Tough Decisions: Drugs and Alcohol" is excellent. Another chapter is "Parents and Families—Can't Live with 'Em, Can't Shoot 'Em." The last chapter lists sources of information on topics of concern and is a valuable resource.

Spangle, Howard. **Voices of Conflict: Teenagers Themselves.** New York: Adama Books, 1987. 272p.

This book deals with the entire range of conflicts experienced by teens. It was written by teenagers who are all students in Glenbard East High School, Lombard, Illinois. Conflict topics range from appearance and self-image, to being gay, to gang activity.

Nonprint Materials

Adolescent Responsibilities: Craig and Mark

Type: 16mm film; color
Length: 28 min.
Cost: Purchase $385
Distributor: Encyclopedia Britannica Educational Corp.
310 S. Michigan Avenue
Chicago, IL 60604
(800) 554-9862
Date: 1973

The development of two teenagers, Craig and Mark, and the influence of their family. Featured are the dynamics as the family discusses future plans—whether or not to leave their familiar location for a new life in the West. The potential sources of family happiness, as well as responsibilities and privileges within the family, receive considerable attention.

Bad Girls

Type: Video; color
Length: 30 min.
Cost: Rental $75, purchase $250

Distributor: Coronet/MTI Film & Video
108 Wilmot Road
Deerfield, IL 60015
(800) 621-2131; (708) 940-3600
(call collect from Illinois and Alaska)
Date: 1990

In the late 1980s, the arrest rate for girls increased ten times faster than for boys. In this film, Deborah Norville interviews a number of female teens who have gotten in trouble with the law for a variety of crimes. Originally produced by NBC News, this film looks at the environment and motives of these teenagers who are in trouble.

Bad Guys, Good Guys

Type: 16mm film; color
Length: 26 min.
Cost: Shipping costs only
Rent from: Indiana University
Audio Visual Center
Bloomington, IN 47405
(800) 552-8620 for cost and arrangements
Date: 1975

This film shows what harm results when violent bullies intimidate other students. Those who are law-abiding and respect the rights of others are easily victimized by the bullies. A special point is made about how ineffective police protection results from poor levels of citizen cooperation.

Chaos to Calm

Type: 16mm film, video; color
Length: 25 min.
Cost: Rental $50 (one week)(film); purchase $450.00 (film), $340 (video)
Distributor: AIMS Media Inc.
6901 Woodley Avenue
Van Nuys, CA 91202
Date: 1985

This program focuses on violent crime, featuring the views of George McKenna, a high-school principal who believes in constructive discipline and student involvement in enforcement.

Delinquency: The Process Begins

Type: 3/4″ video
Length: 28 min.
Cost: Available by rental only
Distributor: Coronet/MTI Film & Video
108 Wilmot Road
Deerfield, IL 60015
(800) 621-2131; (708) 940-3600
(call collect from Illinois and Alaska)
Rent from: Wayne State University
Media Services
5265 Cass Avenue
Detroit, MI 48202
(313) 577-1980 for cost and arrangements
Date: 1976

Two 13-year-old delinquents, each different from the other in ethnicity and family background, are presented. Numerous youth experts then offer their view of the teenagers' problems and the causes.

Failing To Learn . . . Learning To Fail

Type: 16mm film; color
Length: Reel one, 22 min.; reel two, 30 min.
Cost: Rental $15.00 (per reel)
Distributor: Films Incorporated
5547 N. Ravenswood
Chicago, IL 60657
Date: 1977

The case histories of four students show the relationship between juvenile crime and learning disabilities. Though not all those who are learning-disabled become delinquent, the two are shown to be clinically related.

Hard Climb

Type: 16mm film, video; color
Length: 27 min.
Cost: Rental $55 (one week, film or video); purchase $450 (film), $400 (video)

Distributor: Perennial Education
430 Pitner Avenue
Evanston, IL 60202
Rent from: University of Missouri
Academic Support Center
505 E. Stewart Road
Columbia, MO 65211
(314) 882-3601 for cost and arrangements
Date: 1981

A discussion of what it is to be a man takes place between two teenaged boys and an adult as they go on a mountain climbing expedition. They talk about marriage and permanent relationships as opposed to casual sex, and how young men can prove their masculinity without becoming sexually promiscuous.

The Horrible Honchos
Type: VHS video; color
Length: 31 min.
Cost: Rental $49.50
Distributor: Ambrose Video Publishing Inc.
Dept. 1088 C
381 Park Avenue South, Suite 1601
New York, NY 10157-0926
(800) 526-4663
Date: 1977

A gang of kids harasses a new boy in the neighborhood who feels lonely and disliked. This film is based upon Emily Neville's novel *Seventeenth Street Gang.*

Kids in Crisis
Type: VHS or Beta video; color
Length: 28 min.
Cost: Rental $75 (one day); purchase $150
Distributor: Films for the Humanities & Sciences, Inc.
P.O. Box 2053
Princeton, NJ 08543
(800) 257-5126
Date: 1988

Profiles of five very troubled teenagers in this video portray a very frank and realistic examination of the emotional turmoil each of them faces.

Nobody Tells Me What To Do

Type: 16mm film, various video; color
Length: 24 min.
Cost: Rental $50 (five days); purchase $530 (film), $370 (video)
Distributor: Barr Films
P.O. Box 7878
Irwindale, CA 91107
(800) 234-7878
Date: 1984

Gary envies the popularity of several teenage acquaintances until his social activity with them reveals that they have been involved in criminal acts.

Preventing Delinquency: The Social Developmental Approach

Type: 16mm film, various video; color
Length: 28 min.
Cost: Rental $55 (one day); purchase $500 (film), $295 (video)
Distributor: Filmakers Library
124 East 40th Street
New York, NY 10016
(212) 808-4980
Date: 1984

This film offers positive suggestions for dealing with alienation and reducing teen crime. There are many practical suggestions for teachers and parents.

The Reluctant Delinquent

Type: Various video
Length: 24 min.
Cost: Rental $50 (five days); purchase $405

Distributor: Lawren Productions
930 Pitner Avenue
Evanston, IL 60202
(708) 328-6700
Rent from: Indiana University
Audio Visual Center
Bloomington, IN 47405
(812) 335-2103 for cost and arrangements
Date: 1977

A look at Robbie, a 17-year-old in maximum security, and the delinquency that landed him there. Robbie's dyslexia, which created problems for him at school, was a major factor. However, as this film shows, delinquency is much more than a learning disability.

Soapbox with Tom Cottle

Type: Various video; color
Length: 30 min. each
Cost: Rental $95 each (seven days); purchase $150 each
Distributor: PBS Video
1320 Braddock Place
Alexandria, VA 22314
(800) 344-3337
Date: 1985, 1986, and 1987

This excellent series, which includes the titles *Daddy Is 17*, *I Can't Cope*, *I Hate My Body*, *Teenage Depression*, and others, features teenagers being interviewed by psychologist Tom Cottle. Not all individual programs focus on alienated teenagers, but all deal with legal and emotional problems faced by teens.

Stephanie

Type: 16mm, various video; color, b/w
Length: 58 min.
Cost: Rental $100 (film), $85 (video)(plus $15 shipping); purchase $800 (film), $295 (video)

Distributor: Women Make Movies
255 Lafayette St. 212
New York, NY 10012
(212) 925-0606
Date: 1986

Stephanie, a 12th-grader, is a beguiling adolescent who must deal with all of the usual social pressures that face modern teens. This relatively new, award-winning film features in-depth interviews with Stephanie in which she describes her loss of interest in school, despite her obvious intelligence and creativity.

Teen Times: Neither Fish nor Fowl
Type: Various video; color
Length: 28 min.
Cost: Available by rental only
Rent from: Portland State University
Film Library
Portland, OR 97270
(503) 229-4890 for cost and arrangements
Date: 1977

This video describes the dynamics of adolescence and is useful to anyone wanting to understand teenagers better.

Teenage Revolution
Type: 16mm film (2 reels); color, b/w
Length: 50 min.
Cost: Available by rental only
Rent from: University of Iowa
Audio Visual Center
C-5 Seashore Hall
Iowa City, IA 52242
(319) 353-5885 for cost and arrangements
Date: 1965

Produced during the middle of the flourishing counterculture of the 1960s, this film examines the reasons why adolescents feel isolated and apart from the adult generation. They are shown here as idealistic, though naive; somewhat troubled, yet creative and sensitive. Viewers get a good introduction to the youth culture of the "flower-child" era.

Teenagers Talk: Getting through Adolescence

Type: 16mm film, video; color
Length: 12 min.
Cost: Rental $170 ($36 one day, video); purchase $290 (film or video)
Distributor: BFA Educational Media
Phoenix Films
468 Park Avenue South
New York, NY 10016
(800) 221-1274
Rent from: Indiana University
Audio Visual Center
Bloomington, IN 47405
(800) 552-8620 for cost and arrangements
Date: 1975

Provides an overall understanding of adolescent development. From interviews with teenagers, interesting and expressive interpretations of adolescence emerge.

Teens in Turmoil

Type: VHS or Beta video; color
Length: 26 min.
Cost: Rental $75 (one day); purchase $149 (VHS)
Distributor: Films for the Humanities
P.O. Box 2053
Princeton, NJ 08543
(800) 257-5126
Date: 1988

An examination of the crises faced by today's teens—what it's like to be a teenager in America today.

Violent Youth: The Un-Met Challenge

Type: Various video; color
Length: 23 min.
Rent from: Indiana University
Audio Visual Center
Bloomington, IN 47405
(800) 552-8620 for cost and arrangements
Date: 1979

John and William, who are in a juvenile detention home, are interviewed about their past behavior and aspirations for the future. The film deals with violent juvenile crime and shows the viewer the process of the juvenile justice system from detention to punishment.

Organizations

Association of Child Advocates
P.O. Box 5873
Cleveland, OH 44101
(216) 881-2225
President: James J. Lardie

Information exchange on child welfare, juvenile justice, education, and public policy about juveniles.

PUBLICATIONS: Various brochures.

Big Brothers/Big Sisters of America
230 North 13th Street
Philadelphia, PA 19107
(215) 567-7000
Executive Director: Thomas McKenna

Organizations throughout the United States match children from single-parent homes with adult friends who provide guidance and friendship.

PUBLICATION: *Big Brothers/Big Sisters of America Correspondent,* 3 issues per year.

To the Best of You
Six North Michigan Avenue, Suite 1111
Chicago, IL 60601
Founder: Carolyn Shelton

Seeks to provide minority youth with activities and programs for self-improvement and motivation.

PUBLICATION: *From Chitlings to Caviar.*

Volunteers of America (VOA)
3813 North Causeway Boulevard
Metairie, LA 70002
(504) 837-2651

President: Raymond Tremont

Provides more than 400 programs in 200 communities. These include group homes for teens, antidrug programs, and drop-in centers.

PUBLICATION: *Gazette*, quarterly.

CHAPTER 2

Youth Gangs and Violence

LeRoy had seen three members of the Wolves mugging old Mrs. Carson, who lived in the same housing project as he.

> "I'm not going to say anything," LeRoy said.
>
> "I *know* you're not," B.J. said. "Because you're smart, LeRoy. You went and finished school. What I'm saying, LeRoy, is that you're smart enough to know that if you did say something, you'd likely get cut yourself. You know what I mean, LeRoy, when I say *cut*?" Then he grabbed LeRoy's jacket again. This time he didn't slice off a button. He jabbed the point of his knife into the leather and ripped upward. He cut a slash five inches long in the jacket.
>
> If the other three weren't there, LeRoy thought angrily, I'd make him eat that blade. But the other three Wolves were there, and if he started something, he'd wind up getting hurt.
>
> "That's what I mean when I say *cut*, LeRoy," B.J. said.
>
> The Wolves laughed. LeRoy realized he was shaking with anger.
>
> W. E. Butterworth, *LeRoy and the Old Man* (New York: Four Winds, 1980), 8.

Gangs had already ruined LeRoy's Chicago neighborhood. Being caught between gang members' threats and his wish to report truthfully to the police what he had seen was a real dilemma. The agony associated with the decision and the overwhelming fear of the gang are the basis for W. E. Butterworth's excellent book. No easy solutions are available to LeRoy or his neighborhood.

Statistics about Gangs

- Membership in Los Angeles gangs is estimated to be 70,000 (*USA Today*, 1A)
- In 1987, gang-related homicides rose to 387 in the Los Angeles area, and there were 12,000 gang-related arrests (Morganthau, 20)
- Since 1984, Los Angeles gang violence has killed 2,113 people; in 1989 it killed one person every 17 hours (*USA Today*, 1A)
- In Chicago, gang members assessed merchants a "street tax" for insurance while younger gang members collected "tax" from schoolchildren and "tolls" in the elevators of public housing areas (Starr, 32)
- In Martinsburg, West Virginia, a town of only 13,000, Jamaican youth gangs moved in, and 20 gang-related homicides were reported in an 18-month period (Miller, 24)
- At least 150 robberies in a white, middle-class Chicago neighborhood were attributed to the Simon City Royals, who operated in a fleet of stolen four-wheel-drive vehicles (DeMott, 34)

Definitions

Gang. A "group of people who interact at a high rate among themselves to the exclusion of other groups, have a group name, claim a neighborhood or other territory, and engage in criminal and other anti-social behavior on a regular basis" (Hochhaus and Sousa, 74).

Violent gang. "Primarily organized for emotional gratification, and violence is the theme around which all activities center . . .

even delinquent activities are side issues to its primary assaultive pattern" (Yablonsky and Haskell, 262).

Scope of the Gang Problem

On a hot summer evening on Denver's largely black northeast side, a car pulled toward the curb and gunfire ended the life of 16-year-old Tay Carolina, who had been walking down the sidewalk toward his home. The community was filled with rumors about open warfare between the Crips and the Bloods, both of whom had been active in criminal behavior in the neighborhood. The following night 17-year-old Rashid Abdul Riley was shot by a policeman in the same neighborhood after allegedly pointing a loaded .45 caliber handgun at the officer. While the community mourned the loss of the young men, new cries were heard for a crackdown on Denver's gangs, and some concerns were expressed about the lack of recreation opportunities and jobs for young people in the largely minority-populated neighborhood (Robinson, 1).

Are gangs a phenomenon of minority youth who grow up in impoverished neighborhoods? Is the lack of money and economic opportunity the principal cause of young people's seeking out membership in youth gangs? The evidence suggests that the answer is "no" or "not always." A few years ago Mark Miller, a promising student and top high school athlete, was gunned down outside a teen club in an affluent San Fernando Valley, California, suburb. Mark had been a leader in the FFF, a white suburban gang made up primarily of youth from middle-class and wealthy families. For the most part, FFF members were popular at school and often active in school or neighborhood youth activities. Their code was: "(1) Be yourself, (2) live your own life, (3) fuck social values, and (4) fight for freedom" (Sullivan, 52). Hostile and nihilistic, but hardly underprivileged! The gang's purpose for existence appeared to be a selfish, senseless struggle against adult society with its conventional values. Theft for subsistence or to traffic in drugs appeared not to be a primary factor in this youth gang. But gangs such as this may well be the exception.

Gangs exist across the entire United States—from Miami and New York City to the West Coast. They appeal to youth who are reared in poverty. A black sociologist commenting on East Harlem's gangs sees them as alienated youth—society's dropouts. "They are refusing to learn. They are refusing to obey authority. They are refusing to become part of society for all sorts of reasons—mostly negative" (Kolata, C-1, C-13).

In many large cities, gangs are organized largely along ethnic lines (Morganthau, 20). In Los Angeles their principal identity is Chicano, black, or Asian-American, and within the gangs there is a distinctive subculture. In these cases one might view gangs as a more or less natural grouping of ethnically similar youth whose goal is to protect themselves and their neighborhood from the violence brought onto their "turf" by outsiders (who are usually other gangs). Thus they are seen as protective clubs, created of necessity by external social or political circumstances. Some see gangs as the end result of a young person's natural need to belong to a group of others with similar values and characteristics in a hostile and hopeless society.

Although females are sometimes affiliated with gangs, the gang remains primarily a male group. One authority has determined, based upon a national survey, that females account for no more than 10 percent of gang membership, and their roles are usually secondary and supportive (Campbell, 5).

A U.S. Department of Justice report sums it up best: "Hundreds of gangs and thousands of gang members frequent the streets, buildings, and public facilities of major cities; whole communities are terrorized by the intensity and ubiquity of gang violence; many urban schools are in effect in a state of occupation by gangs, with teachers and students exploited and intimidated . . . " (U.S. Department of Justice, 7).

Why Youths Join Gangs

COMPANIONSHIP

Involvement in a group whose members share common values and goals and approve of yours is a direct method of maintaining and enhancing the self-image. Group membership leads to recognition and approval by others, and thereby to an increased sense of self-worth. Therefore, it can be readily seen that companionship contributes to feelings of worth and respect. Many psychologists view companionship as a principal reason that youths seek membership in groups. "Companionship is a critical issue; members want more and closer friends" (Hochhaus and Sousa, 75). Friends give support and endorsement to who we are and what we are all about. The sense of identity and the self-worth of the adolescent are validated and enhanced. Gang activity, even though it may be illegal, often fulfills the basic need asked of companionship—the need to be esteemed by one's fellows.

> "Any act, violent or nonviolent, which defies the social and legal authority of the larger society may increase the status of the gang member in the eyes of his peers. Furthermore, the greater the risk of danger to himself, as well as to others, the more masculine and powerful he may seem to his peers, thereby enhancing his sense of worth and self-esteem."
> (Friedman, Mann, and Friedman, 601).

In a study of more than 500 gang members in Philadelphia, psychologists found that companionship ranked among the highest of the benefits that gang members thought they had gained from being in gangs. The study further found that not only do gang members express less guilt over violence and theft than do nonmembers, but also that once a nonmember joins a gang, he quickly assumes the antisocial values of those already in the gang (ibid., 600–603).

PROTECTION

The need for physical protection from harm is another frequently mentioned reason for joining a gang. There are neighborhoods and schools where any person is a potential victim of unprovoked violence. It is not just in the ghetto, nor is it just in instances where those victims of violence have been reckless about personal safety or somehow antagonized the perpetrators of violence. "Protection attracts young people to gangs; and to a certain degree, they are protected. If a member is threatened or victimized by a rival gang or any other individual, his gang will usually come to his defense" (Hochhaus and Sousa, 75).

Though self-defense appears to be a fully justifiable behavior, the motive is also used as an excuse to harass other gangs in what would appear to be a continuous battle to "defend the turf." Thus considerable gang time and energy is spent on trying to convince rival gangs to leave the neighborhood, or at least to acknowledge the dominance of the other. The struggles seem endless; the battles are often bloody. Frequently deaths of gang members result from this tit-for-tat behavior of the gangs, and almost always the reason given is to "protect the turf." "It is important to note that while a gang as a whole promotes the violence, the individual members may oppose it; they go along as a result of peer pressure" (ibid.).

PEER PRESSURE

Peer pressure is not limited to adolescents and young adults, but because adolescence is a time of transition between childhood and young adulthood, peer pressure may be felt with greater intensity. Peer pressure influences behavior in the classroom, choices of social activities, food and clothing, hair styles, music, and speech. The need to get along with others is basic, and leads to adaptation. To want to conform is natural and almost universal. One authority puts it this way: "One way the individual has of being a part of a particular group is to be like other members of the group. . . . When a fad is in fashion, every person in the group adopts it. Those who are different are excluded as 'queer' or 'gross'" (Rice, 303). The problems that arise from conformity to the group exist when the basic values of the group are antisocial, and crime and violence are an accepted norm. When the group to which one belongs is a gang, then each member's behavior is likely to conform to that antisocial norm.

REFERENCES

Barich, Bill. "The Crazy Life," *The New Yorker* 62, no. 37 (3 November 1986): 97–130.

Bosc, Michael. "Street Gangs No Longer Just a Big-City Problem," *U.S. News & World Report* 97, no. 3 (16 July 1984): 108–109.

Butterworth, W. E. *LeRoy and the Old Man*. New York: Four Winds, 1980.

Campbell, Anne. *The Girls in the Gang*. Cambridge, MA: Basil Blackwell, 1984.

DeMott, John. "Have Gang, Will Travel," *Time* 126, no. 23 (9 December 1985): 34.

Friedman, C. Jack, Fredrica Mann, and Alfred Friedman. "A Profile of Juvenile Street Gang Members," *Adolescence* 10, no. 40 (Winter 1975): 563–607.

Hackett, George. "Kids: Deadly Force," *Newsweek* 61, no. 2 (11 January 1988): 18–19.

Hochhaus, Craig, and Frank Sousa. "Why Children Belong to Gangs: A Comparison of Expectations and Reality," *High School Journal* 10, no. 1 (December 1987/January 1988): 74–77.

Kolata, Gina. "Grim Seeds of Park Rampage Found in East Harlem Streets," *New York Times*, 2 May 1989: C-1, C-13.

Miller, Mark. "A Jamaican Invasion in West Virginia," *Newsweek* 111, no. 13 (28 March 1988): 24.

Morganthau, Tom. "The Drug Gangs," *Newsweek* 111, no. 13 (28 March 1988): 20–27.

Muehlbauer, Gene, and Laura Dodder. *The Losers: Gang Delinquency in an American Suburb*. New York: Praeger, 1983.

Rice, F. Phillip. *The Adolescent: Development, Relationships and Culture*. 4th ed. Boston: Allyn & Bacon, 1975.

Robinson, Marilyn. *Denver Post*, 16 July 1988: 1.

Seng, Magnus. "Gang Development in Suburban Communities: The Role of the High School," *American Secondary Education* 15, no. 1 (1986): 12–16.

Starr, Mark. "Chicago's Gang Welfare," *Newsweek* 35, no. 10 (28 January 1985): 35.

Sullivan, Randall. "Leader of the Pack," *Rolling Stone*, no. 481 (28 August 1986): 51–84.

U.S. Department of Justice, Law Enforcement Assistance Administration. *Violence by Youth Gangs and Youth Groups in*

Major American Cities. Washington, DC: U.S. Government Printing Office, 1976.

USA Today, 7 December 1989: 1A.

Yablonsky, Lewis, and Martin Haskell. *Juvenile Delinquency*. 4th ed. New York: Harper & Row, 1988.

Resources
for Finding Out about Youth Gangs and Violence

Fiction

Bonham, Frank. **Durango Street.** New York: Dutton, 1965. 190p.

The Gassers are the gang that has dominated Durango Street for a long time. Rufus Henry discovers that there is no escape from their violence, despite police surveillance, because the Gassers operate singly or in small unnoticed groups, not as a pack. When he discovers that he can't escape, he is forced to join another gang, the Moors, in order to survive. Although this is a violation of his parole, Rufus soon fights his way to leadership of the Moors. The story follows Rufus to a final showdown when the Moors hold a party at Club Chic Hall.

Bowman, Frank. **Cool Cat.** New York: Dutton, 1971. 151p.

While the Machete gang deals in dope and violence in their neighborhood, Buddy Williams and his friends Rich and Cool buy an old pickup truck and go into the hauling business. Whenever there is trouble with the Machetes, a cool cat named Cal Brown seems to be hanging around. Though never directly involved in the problems, Brown isn't the kind of guy Buddy could like. At the end of this story of ghetto gangs and straight kids, the role of the "cool cat" becomes apparent.

Butterworth, W. E. **LeRoy and the Old Man.** New York: Four Winds, 1982. 154p.

When LeRoy comes home to his Chicago housing project, he becomes a witness to the brutal beating and robbery of an old woman neighbor. The Wolves gang warns him not to expose them to the police, yet LeRoy is troubled by the violence he saw. He flees to his grandfather in Mississippi, where he learns a different set of values. Conflict between these values and the rule of the Wolves continues to torment him until he reaches a critical decision.

Cate, Dick. **Flames.** New Pomfret, VT: David and George, 1989. 160p.

Billy learns that gang life is self-defeating, as his friendship for Emma grows. The setting is in northern England, but the dynamics of gang activity are universal.

Forshay-Lunsford, Cin. **Walk through Cold Fire.** New York: Delacorte, 1985. 205p.

Written when the author was 18, this story is about a 16-year-old who leaves her comfortable middle-class home to spend a summer with her aunt. While there she becomes acquainted with the Outlaws and falls in love with Billy. The emotional conflict and the clash of values that she experiences form the body of this novel.

Guy, Rosa. **The Friends.** New York: Holt, Rinehart & Winston, 1973. 203p.

Survival on the streets of New York is the theme of this story about Phyllisia, whose family has moved to Harlem from the West Indies. Because of her accent, her classmates brand her as odd and refuse to be friends. So Phyllisia turns to the only other girl who will have her and becomes a friend to Edith, whose family lives in poverty. Their friendship and struggles for survival form the basis for this book.

Hinton, S. E. **The Outsiders.** New York: Viking, 1967, 160p.

A 17-year-old author describes the struggles of a gang of tough, long-haired kids called the Outsiders against the Socs, a gang made up of privileged kids. This book tells of violence and hatred, but also of the trust and loyalty within the gang. It gives us a look at young people at a crossroads, trying to decide if the life of a hoodlum is what they really want.

Hinton, S. E. **Rumblefish.** New York: Viking, 1976. 122p.

Rusty-James wants to be tough like his older brother, Motorcycle Boy. He wants to be the most feared hood on his side of the river, but, unlike Motorcycle Boy, he doesn't always use his head. Motorcycle Boy comes to his rescue whenever things don't go well, until at one point things really fall apart, and older brother isn't there to help Rusty-James out.

Hinton, S. E. **That Was Then, This Is Now.** New York: Viking, 1971. 224p.

Bryon and Mark are best buddies doing everything together—playing pool, running with the gang, and fighting against the Socs. "That's just the way things are." Bryon is often bothered by things, things that don't make much sense, but Mark always says that it is fate. Then one day Bryon sees what happens to M&M, his girlfriend's little brother, and nothing is the same after that.

Holman, Felice. **Slake's Limbo.** New York: Scribner's, 1974. 117p.

Another story about survival on the streets of New York. Aremis Slake is a motherless 13-year-old who is small for his age, has poor eyes, and is useless to any gang. Picked on by other teens, he finds a tunnel leading from the subway, and survives for four months in his cave beneath the city streets.

Murphy, Jim. **Death Run.** New York: Clarion, Houghton Mifflin, 1982. 174p.

Brian and his buddies play a prank on a boy in a park, but the prank goes wrong and the boy dies. As the police search for the killer, Brian feels fear and guilt. Sgt. Wheeler pursues every lead, and eventually comes face to face with Brian.

Myers, Walter D. **Fast Sam, Cool Clyde, and Stuff.** New York: Viking, 1975. 187p.

When Fast Sam, Gloria, BB, Cool Clyde, and others form a group in friendship, they actually start a gang. Stuffing the ball (in basketball games) is a favorite activity, and a brush with police gives this funny story the flavor of real life in the ghetto.

Platt, Kin. **Headman.** New York: Greenwillow, 1975. 186p.

Violence and bloodshed are the fate of gang members on a Los Angeles street. Owen Kirby's struggle to escape ghetto violence is the subject of this book.

Nonfiction

BOOKS

Bernstein, Saul. **Alternatives to Violence.** New York: Association Press, 1967. 192p.

An excellent source of information about the impact of ghetto culture on adolescents, and the attraction of gangs even today.

Bernstein, Saul. **Youth on the Streets.** New York: Association Press, 1964. 160p.

Although this book is more than 25 years old, it contains many details on the composition, dynamics, and characteristics of youth gangs that still hold true. Of particular interest is Chapter 5, which discusses a variety of methods of working with alienated street gang members.

Brown, Waln K. **The Other Side of Delinquency.** New Brunswick, NJ: Rutgers University Press, 1983. 188p.

An autobiographical account of a life of delinquency by a man who is now a successful professional. His description of life in reform school is particularly interesting.

Miller, Alden E., and Lloyd E. Ohlin. **Delinquency and Community.** Beverly Hills, CA: Sage, 1985. 208p.

This book is a discussion of theory and research findings. Chapter 6 on "Community Change, Social Control and Juvenile Delinquency" contains findings with serious implications for school-based gangs.

Morris, Edward K., and Curtis Braukmann, eds. **Behavioral Approaches to Crime and Delinquency.** New York: Plenum Press, 1987. 625p.

A resource on many aspects of juvenile crime. Particularly recommended is the discussion of several approaches to gang activity.

Muehlbauer, Gene, and Laura Dodder. **The Losers: Gang Delinquency in an American Suburb.** New York: Praeger, 1983. 138p.

One of the best studies of gang activity available. Chapter 4, "The Gang Forms," provides good insight into the manner in which new group values take over those of individual gang members.

Roberts, Harrell B. **The Inner World of the Black Juvenile Delinquent: Three Case Studies.** Hillsdale, NJ: Lawrence Erlbaum Associates, 1987. 140p.

Excellent case study material on three black youths, including their gang activity. The case of Sly and his involvement with a street gang makes compelling reading.

Sutton, John. **Stubborn Children.** Berkeley, CA: University of California Press, 1988. 298p.

This is both a historical and a theoretical look at American child control. The book offers research and theory on the stubborn or troublesome child and the implications for involvement in gangs.

U.S. Department of Health, Education, and Welfare. **Violent Schools—Safe Schools: The Safe School Study Report to the Congress.** Washington, DC: U.S. Government Printing Office, 1977. 14p.

This is a study of crime and disruption in schools, indicating a high correlation between crime within the school and fighting gangs in the school's neighborhood.

U.S. Department of Justice, Law Enforcement Assistance Administration. **Violence by Youth Gangs and Youth Groups in Major American Cities.** Washington, DC: U.S. Government Printing Office, 1976. 79p.

A summary of a research report drawn from work done by Harvard Law School for the government. Although the findings are now dated, this report on gangs in six large cities has implications for society today.

Webb, Margot. **Coping with Street Gangs.** New York: Rosen Publishing, 1990. 150p.

A good guide to what street gangs are all about, this book provides answers to why gangs exist, why teens join gangs, what the inner workings of gangs are, and why there is so much violence associated with them. The author also adds insights into how gangs affect family members, and where to go for help.

ARTICLES

Barich, Bill. **"The Crazy Life,"** *New Yorker* 62, no. 37 (3 November 1986): 97–130.

This is a very detailed account of gang activity and the lives of gang members in Los Angeles County.

Bosc, Michael, Steve Hawkins, and George White. **"Street Gangs No Longer Just a Big City Problem,"** *U.S. News & World Report* 97, no. 3 (16 July 1984): 108–109.

Small towns and the suburbs are now also feeling the effects of youth gangs and their crimes. This article points out that the high degree of violence inherent in gang activity is a nationwide problem.

Bowker, Lee H., Helen Gross, and Malcolm Klein. **"Female Participation in Delinquent Gang Activities,"** *Adolescence* 15, no. 59 (Fall 1980): 509–519.

Research data on the effect of females on gang activity. Most often, males try to exclude females because their presence is inhibiting.

DeMott, John. **"Have Gang, Will Travel,"** *Time* 126, no. 23 (9 December 1985): 34.

The mobility of gangs and the spread of their violence outside the ghetto is the focus of this report.

Friedman, C. Jack, Fredrica Mann, and Howard Adelman. **"Juvenile Street Gangs: The Victimization of Youth,"** *Adolescence* 11, no. 44 (Winter 1976): 527–533.

These researchers studied gang members in Philadelphia to determine the nature and extent of their violence, and the degree of gang influence on and control over individual members' behavior.

Gest, Ted, and Cynthia Kyle. **"Kids, Crime, and Punishment,"** *U.S. News & World Report* 103, no. 8 (24 August 1987): 50–51.

The authors describe the national picture on youth crime and gangs, and present approaches to the problems by authorities in different areas.

Hackett, George, and Michael Lerner. **"L.A. Law: Gangs and Crack,"** *Newsweek*, 109, no. 17 (27 April 1987): 35–36.

Battling for market share in the cocaine trade, the Bloods and Crips gangs are virtually in control of parts of Los Angeles. Youths as young as 11 play a big part in the rich crack market, and share the accompanying violence.

Hackett, George, et al. **"Saying 'No' to Crack Gangs,"** *Newsweek* 61, no. 13 (28 March 1988): 29.

Los Angeles gangs are facing community resistance, and in Brooklyn the crack gangs are being countered by organized Muslim activity.

Hackett, George, Richard Sandaz, Frank Gidney, Jr., and Robin Gareiss. **"Kids: Deadly Force,"** *Newsweek* 61, no. 2 (11 January 1988): 18–19.

This is a description of the scope of teenage violence, much of it related to drug trade and gangs.

Hochhaus, Craig, and Frank Sousa. **"Why Children Belong to Gangs: A Comparison of Expectations and Reality,"** *High School Journal* 71, no. 2 (December 1987/January 1988): 74–77.

This is an excellent review of the psychological motives for the involvement of young people in gangs.

Lowney, Jeremiah. **"The Wall Gang: A Study of Interpersonal Process and Deviance among Twenty-Three Middle Class Youths,"** *Adolescence* 19, no. 75 (Fall 1984): 527–537.

Research on middle-class teenagers in gangs shows that they feel alienated and express high levels of distrust of all adults.

Morganthau, Tom, et al. **"The Drug Gangs,"** *Newsweek* 61, no. 13 (28 March 1988): 20–27.

All across this country, gangs that include hundreds of teenagers are involved in drug traffic. Violence, including homicide, often characterizes the turf wars.

Seng, Magnus. **"Gang Development in Suburban Communities: The Role of the High School,"** *American Secondary Education* 15, no. 1 (1986): 12–16.

This article describes the stages of gang development, beginning with adolescent alienation, through organized criminal activity.

Starr, Mark, and Frank Maier. **"Chicago's Gang Warfare,"** *Newsweek* 35, no. 10 (28 January 1985): 32.

Teenage gangs rule parts of Chicago with terror, extorting "street tax" from some and business protection insurance from others.

Nonprint Materials

The Gang

Type:	1/2" video
Length:	15 min.
Cost:	Rental $50 (one week; 30-day free preview, pay shipping only); purchase $300 (video)
Distributor:	FMS Productions P.O. Box 4428 520 E. Montecito, Suite F Santa Barbara, CA 93140 (800) 421-4609
Date:	1977

After teenagers vandalize a school, the community gathers to blame and share responsibility.

Gangs: Not My Kid

Type:	Video, color
Length:	48 min.
Cost:	Rental $75; purchase $250
Distributor:	Coronet/MTI Film & Video 108 Wilmot Road Deerfield, IL 60015 (800) 621-2131; (708) 940-3600 (call collect from Illinois and Alaska)
Date:	1990

Youth gangs and their violence create a negative impact on family life, and this documentary focuses on the evolution of a gang member in a tough Los Angeles neighborhood. Tyne Daly hosts this film, which won the CINE Golden Eagle award.

Gangs: The Consequences of Conformity

Type:	16mm film, various video
Length:	16 min.
Cost:	Purchase $300 (film), $110 (video)
Distributor:	AIMS Media Inc. 6901 Woodley Avenue Van Nuys, CA 91201 (818) 785-4111
Rent from:	University of Wisconsin Bureau of Audio-Visual Instruction 1327 University Avenue P.O. Box 2093 Madison, WI 53701-2093 (608) 262-3402 (for cost and arrangements)
Date:	1977

Conformity and membership in gangs is explored in detail. In some groups there is a high level of pressure to conform to group standards, and the results are often negative and antisocial.

Gangs, Cops, and Drugs (Parts I and II)

Type:	Video, color
Length:	49 min. (each part)
Cost:	Rental $75 (each part); purchase $250 (each part)

Distributor: Coronet/MTI Film & Video
108 Wilmot Road
Deerfield, IL 60015
(800) 621-2131; (708) 940-3600
(call collect from Illinois and Alaska)

Date: 1990

This up-to-date film focuses on the gang problem in Los Angeles with its 700 gangs. Produced by NBC News and hosted by Tom Brokaw, both parts show the devastating consequences of youth gang crime. Part II deals particularly with the combination of gang membership and drugs.

Organizations

Guardian Angels
982 E. 89th Street
Brooklyn, NY 11236
(718) 649-2607
Founder: Curtis Sliwa

The Guardian Angels seeks to provide positive role models for youth. Volunteers patrol streets and inner city communities to combat gangs and prevent crime.
PUBLICATION: *Streetsmarts: The Guardian Angels' Guide to Safe Living.*

Youth Emotions Anonymous
P.O. Box 4245
St. Paul, MN 55104
(612) 647-9712
Coordinator: William Roath

The group provides support and encouragement for youths ages 13 to 18 who want to become healthier emotionally, and refers inquirers to local support groups for teenagers with low self-esteem.

Youth Organizations U.S.A.
P.O. Box 202
Teaneck, NJ 07666

President: Julian I. Garfield

This organization provides minority youth with programs designed to promote positive values and self-awareness, and produces "YOUSA Speaks," a cable television program.

CHAPTER 3

Substance Abuse

> Nick went to the liquor cabinet, partly to see if it would work. He found the whiskey bottle, he got a glass out of the cabinet, and he put some whiskey in it. . . . The whiskey really did taste ugly, it burned his throat, and pretty soon he was sleepy.

In Virginia Wolff's novel, *Probably Still Nick Swansen,* a first attempt at drinking to feel good leaves 16-year-old Nick feeling worse. Nick Swansen had never tried booze before, but he had seen a movie in which a man had drunk some whiskey in order to feel good. Nick was hurting. He felt empty and lonely. He was 16, a learning-disabled guy, which meant to him that everyone else was smarter than he.

> Nick continued to drink whiskey, doze, and watch TV most of the afternoon until finally . . . he was sick to his stomach, and he woke up. He tried to make it to the bathroom, but his stomach beat him in the race. He was only partway out of the door of his bedroom when he threw up.
>
> He couldn't stop throwing up, and then he couldn't throw up enough. Please don't let Mom and Dad get home now, he said to the floor.
>
> Virginia Euwer Wolff, *Probably Still Nick Swansen* (New York: Henry Holt, 1988), 73–74.

> Rob found matches in Fairlee's glove compartment. It was raining steadily now, a rhythmic pound on the roof. The windows fogged up. He locked the doors

> and breathed in the first curl of smoke. His body shuddered with it. God, those boys knew what they were doing. This was the best dope he'd ever had. "Better than you, Ellery Collier," he said aloud, and sank into himself willing the calm to gather in his head, to ooze like thick warm syrup into his fingertips, his toes, his stomach. He was surprised how fast the joint went.
>
> . . . He pressed the roach out in the ashtray and put it back in the bag. His head was swimming and his eyes itched slowly. You're not buzzed, he thought to himself. What you need, boy, is a cloud right in the center of your brain, a soft foamy cloud you can sleep in. He hadn't slept all night. . . . He remembered the other joint. . . . He sucked it in, held the smoke, felt it spiraling down, loose in his chest.
>
> Like an angel, he thought, feeling a soft buzz behind his eyes. He pulled again, pressing his lips tight to the warmth. It was like kissing an angel. . . .
>
> In his stupor, Rob missed the curve in the road. His misery, being arrested, and going on trial later followed his high. He was no longer kissing an angel.
>
> Sue Ellen Bridgers, *Permanent Connections* (New York: Harper & Row, 1987), 171–172.

Pot is popular, cheap, and readily available. It was the "hip" drug of the 1960s and 1970s, used by more teens over the past ten years than any drug other than alcohol (Johnston, O'Malley, and Bachman, 48).

Statistics about Substance Abuse

- When asked about alcohol use during the preceding year, 42 percent of eighth graders and 53 percent of ninth graders acknowledged drinking once to ten times or more (Benson, Williams, and Johnson, 144)

- Twenty percent of ninth graders reported that they had used marijuana during the previous year (ibid., 146)
- Sixty percent of high school seniors in 1989 said that they had used alcohol during the preceding 30 days ("National High School Senior Survey," 5)
- Cocaine use among high school seniors dropped from 6.2 percent in 1986 to 2.8 percent in 1989 (ibid., 1)
- Crack use among high school seniors dropped from 3.9 percent in 1987 to 3.1 percent in 1989 (ibid., 2)
- Marijuana use among high school seniors was down from 37 percent in 1979 to 17 percent in 1989 (ibid., 1)
- About 30 percent of high school seniors in 1989 reported that they smoked cigarettes—a rate that was unchanged over the previous eight years (ibid., 3)

Definitions of Substances

Marijuana. From the plant known as cannabis. Usually the leaves are dried, then smoked as cigarettes. Also known as pot or grass, its primary chemical agent is THC.

Alcohol. Any beverage, including beer, wine, whiskey, vodka, gin, and liquors, that contains alcohol. Also referred to as booze or juice, it is legally sold to those 21 years or older, in most states, in specially licensed establishments.

Hashish. Also called hash. Made from the resin of the cannabis plant, its primary chemical agent is THC, and it is somewhat more potent than marijuana.

Cocaine. A central nervous system stimulant made from the leaves of the coca plant. Known in various forms as crack, rock, or freebase rocks, it is a highly addictive stimulant.

Crack. A form of cocaine that has been converted to a paste that hardens as it dries. It is then cut or broken into chunks that resemble soap or gravel rocks, which are usually smoked rather than sniffed.

Amphetamines and methamphetamines. Central nervous system stimulants that can be smoked, swallowed, or injected. Include bennies, dexies, speed, wake-ups, and ice. These are synthetic drugs legally made and sold by the pharmaceutical industry, but they can also be made by illegal basement labs.

Hallucinogens. Substances whose primary chemical agent produces sensations or perceptions that are not based upon reality, ranging from simple fantasies to psychotic episodes.

Heroin. Derived from the poppy plant, heroin is usually injected. Also known as horse, smack, junk, or scag, it is classified as a narcotic.

Scope of the Problem

Anyone who hears a news broadcast or reads a daily newspaper knows that drug and alcohol abuse have become a national problem. Teens as well as adults are seriously affected. The costs to society are enormous, and many experts believe that reducing usage among teens is the most logical place to begin solving the problem. A recent Gallup poll indicated that people believe that drug use is the number one problem in American schools ("Getting Tough," 40).

Most, though not all, substances that are misused are illegal. Alcohol and tobacco are controlled by law and are illegal only for children and teens. The minimum age for the legal purchase of tobacco products varies among the states. Practically all states now require people to be 21 or over in order to legally purchase and use alcoholic beverages. Obviously, these restrictions have been difficult to enforce, and many teens choose to get around the laws one way or another.

ALCOHOL

By far, the most popular of all substances used and abused by teenagers is alcohol. According to the University of Michigan survey, one of the most respected, 92 percent of high school seniors throughout the United States have tried it. In that survey, 85 percent of the high school seniors acknowledged using alcohol within the past 12 months, and 35 percent admitted to having as many as five drinks in a row within the two weeks just prior to the survey ("Teen Drug Use," 6).

First, consider why this is the case. Alcohol is an accepted part of the American way of life. Right or wrong, the use of alcohol is imbedded in our culture. TV commercials glamorize it; countless movies and television shows portray it as a part of our social life. And adults fear it less than other drugs. Parents tend to fear drug use by their teens far more than alcohol use (Groller, 26). For the time being, at least, it appears that parents are far more tolerant of alcohol use than of drug use by their teenagers. They seem to be relieved to know that their kids aren't snorting coke or smoking pot. "Some parents view the issue of teen-age drinking as an exercise in damage control. . . . Many recall their own experimentation with booze and resolve to teach their children how to drink responsibly" (Gibbs, 94).

It seems that even young children don't regard alcohol in the same light as drugs. In a recent survey of fourth graders, 84 percent identified marijuana as a drug, but only 42 percent knew that alcohol is also a drug (Donovan, 189). Though parents may be somewhat tolerant of underage drinking, the justice system doesn't necessarily view it with leniency. A New Jersey judge recently sentenced two Princeton University students found guilty of providing alcohol to minors to 30-day jail sentences and $500 fines ("Campus Dryout," 70). In general, however, most teens expect parents to be less frightened of their use of alcohol than of other drugs.

In addition to its greater acceptance by society, alcohol is readily available. Teens who want to use alcohol can usually find an older friend or brother or sister to buy them a six-pack or a bottle of liquor. Retail outlets are everywhere, and most better restaurants serve alcohol in connection with their food business. Its use is an integral part (according to our literature and films) of gracious dining and adult social activity. And finally, it is present in many homes, easily available to any family member. Dad's liquor cabinet is often left unlocked, and the fun of getting high is within arm's reach.

The use of alcohol is practically as old as the human race. "No one knows what kind of liquor was first, wine or beer or mead; but by the Neolithic Age, it was everywhere" (Kinney and Leaton, 6). Ancient Greek mythology includes references to alcohol as a gift from the gods. Priests drank it. Warriors used it to bolster their courage before battle. The Bible contains several references to the use of alcohol. The New Testament in the book of John describes the miracle in which Christ turned water into wine for a wedding celebration.

The fact that alcohol has been around for so long has permitted the accumulation of considerable scientific knowledge about its harmful physical effects. Other drugs, such as marijuana, may be more or less harmful, according to which research you value most. We simply have more scientific data about the harmful effects of alcohol on the human body than about any other drugs (Barnes, 1729). This may help to explain why some parents worry less about their teen's use of alcohol than of other substances, whose long-range effects are not quite so well understood.

Prolonged use of alcohol causes numerous physical problems. The most widely recognized of these problems is liver disease. Three of these diseases are: (l) fatty liver, (2) alcoholic hepatitis, and (3) cirrhosis. In cirrhosis, liver cells are destroyed and replaced by scar tissue. This damage is irreversible even when drinking stops, and the ultimate result is that the liver cannot do its work properly. There are often complications, such as pancreatitis and cancer, and the liver damage often leads to death.

Among less well known physical effects are reduced production of red blood cells and the decreased effectiveness of white blood cells. Thus the body is less able to cope with other diseases. Other problems associated with alcohol abuse include enlargement of the heart, abnormalities in cardiac rhythm, urinary tract infections, reduced respiration rates, and a variety of stomach and intestinal ailments. Research evidence on effects of alcohol use for more than 10 to 15 years is overwhelmingly negative, even though usage may be thought to be "moderate." All in all, not a pleasant consequence, though alcohol use probably begins in most cases as a simple experimentation with "getting high" (Estes & Heinemann, 19–26).

Injuries and deaths of innocent people through drunk driving are another negative consequence. (Of course, drunk driving is a problem of all Americans, not just teens.) A vivid and tragic example is the case of Bruce Kimball, champion diver and Olympic athlete. He was a college freshman returning from a friend's wedding reception when the car in which he was riding was struck by a drunk driver. Kimball's left leg was broken, the knee shattered, his spleen ruptured, and every bone in his face broken. He then began a long period of recovery and rehabilitation capped by his return to international diving competition. Unfortunately, the story doesn't end happily there. On August 1, 1988, he was driving his car after having a few beers himself when he lost control. His car shot through a crowd of young people, killing two and seriously injuring six others. He was charged with vehicular homicide and has not been a serious international diving competitor since (Plummer, 87–88).

On the good news side, there is some evidence that heavy drinking among teenagers is slacking off. A comprehensive annual survey by University of Michigan social scientists shows alcohol use among high school seniors is in a decline. High school seniors who had drunk within the 30 days prior to the survey dropped from 72 percent in 1980 to 60 percent in 1989 ("National High School Senior Survey," 5). Heavy drinking, defined as five or more drinks in a row during the preceding two weeks, also dropped. Heavy drinking among these high school seniors was down from 41 percent in 1983 to 33 percent in 1989 (ibid.).

MARIJUANA AND HASHISH

Although marijuana is popular and readily available, it is illegal to sell, and in most states it is illegal to possess. Commonly known as pot, it is believed by many to be a light drug with few or no serious long-term physical effects. "Pot's like nothing, you know, you smoke it and you get high, and then that's it. It doesn't really affect me that much anymore," said one teenage user (Glassner and Loughlin, 87). Like alcohol, marijuana has been around since pre-Biblical times, but the long-term physical effects of heavy usage are not quite as well researched, particularly in the area of chromosome abnormalities.

The following conclusion was made by a scholar who recently examined all of the available research evidence on this point:

> It should be clear at the outset that there are no obvious, high-frequency, serious physiological effects of moderate use of marijuana over a 5- to 10-year period. If there were, we would see the results all around us. But then, there are no obvious, high-frequency serious effects of moderate use of cigarettes or alcohol over a 5- to 10-year period.
>
> (Ray, 437)

According to the University of Michigan survey, usage of marijuana and hashish by high school seniors was down from a high of 37 percent in 1979 to 17 percent in 1989. Again, these data refer to use during the 30 days preceding the survey. As for daily use during that same time period, the rate was only 3 percent ("National High School Senior Survey," 10).

COCAINE AND CRACK

Once regarded as the drug most favored by young professionals, cocaine has become one of the most often used drugs of the 1980s. The total number of cocaine users in the United States doubled between 1977 and 1981 (Hazelden Foundation, 5). One drug counselor observed that there are two trends in the use of cocaine: "Younger and younger and more and more" ("Kids and Cocaine," 58). There is some evidence, however, that the rate of cocaine use among teens is dropping off slightly ("National High School Senior Survey," 2). The downward trend in cocaine use by high school seniors has been evident during the past four years, and may represent a significant change in teen drug preference (ibid., 9).

In recent years, the use of crack, a refined form of cocaine, has received considerable media attention. Crack is usually smoked, and is generally believed to be more highly addictive than cocaine, which is snorted. However, the national surveys indicate a relatively small proportion of teenagers using crack. Less than 4 percent of high school seniors reported using crack at any time during the previous year, and the trend of use is slightly downward (ibid., 2).

But the problem remains, especially in the inner cities where poverty abounds. The lure of "get rich quick" has proven to be a quick route into big-time crime for money. The big bucks are there waiting, and teens are in the coke business in a big way. "They are being used to process, package, cut and distribute, to sell and even to enforce discipline in the ranks," said one law officer (Lamar, 22). Children as young as nine or ten enter the business as lookouts, with rewards such as a radio or a leather jacket. Later they become runners, whose job is to distribute coke from wholesaler to dealer. Runners have been known to make as much as $300 per day (ibid.). Often those who work in the cocaine sale and distribution business are not users themselves—they are young entrepreneurs. "Through the mid-1980s, Detroit's major drug entrepreneurs resembled Fortune 500 companies in their size and scope. Police estimated Young Boys Inc., one of the city's most notorious gangs, had revenue of $7.5 million a week at its peak" (Hundley, I-6).

AMPHETAMINES AND METHAMPHETAMINES

Amphetamines produce wakefulness and alertness. Some doctors now prescribe amphetamines for children diagnosed as having attention deficit disorders. Users report feeling good, or a state of well-being. Methamphetamine is a closely related chemical compound with effects quite similar to amphetamine—perhaps a "higher" high. Ice is a crystal methamphetamine with many effects that are similar to crack. Amphetamines can be obtained in liquid form and injected with hypodermic needles. Obviously, this raises the common fear of infection, especially with hepatitis and AIDS. One of the side effects long known is the loss of appetite. Thus, they appeal to those who wish to lose weight.

Long-term physical effects are not entirely clear. Numerous fatalities have been associated with prolonged heavy amphetamine use. In the 1960s it was common to hear that "Meth is Death" and "Speed Kills." Whether deaths resulted directly from the amphetamines or from the infections that followed was not always clear. There is more agreement among experts on the high potential for mental illness. There have been numerous cases of paranoid psychosis reported, and the exact connection with various levels of amphetamines remains under study.

HALLUCINOGENS

Like alcohol and cocaine, some of the hallucinogens have been around for centuries. Ancient Norsemen allegedly consumed mushrooms to produce pleasant fantasies as well as courage for battle. The discovery of lysergic acid compounds (LSD) in the late 1930s led to some early experiments on artificially induced psychoses. Finally the use and resultant publicity concerning psilocybin mushrooms by one-time Harvard psychologist Timothy Leary opened the door for more experiments and misuse (Ray, 234–238). Favored by the hippies of the 1960s, LSD became popular, and American youth were urged to "turn on, tune in, and drop out." Users reported that they experienced a heightened state of awareness, exhilaration, and a variety of insightful experiences. Equally apparent to some users (bad trippers) were visions of horror, nightmares, states of terror, and unprovoked violence. "Flashbacks" or repeat episodes long after the initial use of acid are not uncommon. Similar hallucinations, as well as pleasant changes in images, are reported by mushroom users (ibid.). Although some legitimate research on these substances continues, the long-term effects are unknown.

HEROIN

Heroin has been, in times past, the drug of choice in the ghetto, but its popularity has lessened a bit with the rise of cocaine. The University of Michigan research shows that the percentage of users was less than 1 percent among high school seniors nationwide in 1989 ("Teen Drug Use," 6).

Heroin users usually describe their sensations as a "rush." Not all users become addicted, but the probability is quite high (Ray, 348). Withdrawal is usually painful and difficult. "Whether you smoke it, chip it, shoot it, or drink it—all the narcotics are addicting" (ibid., 349).

TOBACCO

Although tobacco is a legal substance, its use by those who are underage is illegal in most states. The negative health effects are very well known, and have become the focus of periodic reports to the nation by the U.S. surgeon general. Nevertheless, usage among teenagers continues at a substantial rate. The most recent survey by University of Michigan scientists shows that the rate of cigarette smoking among high school seniors is about 30 percent, and has remained unchanged for the past ten years ("National High School Senior Survey," 5).

Why Teens Do Drugs

Many wonder why anybody, especially a teenager, begins using drugs when the harmful effects of long-term use are well known. As with the other behaviors described in this book, the motivation to do drugs is not totally clear. However, the reasons teens give fall into several categories.

PEER PRESSURE

As was discussed earlier, the desire to have the approval of one's peers is almost universal, especially for teens. There is a social need to be with peers and to behave as they do. When the peer standards include drug and alcohol use, most teens feel it necessary to experiment, to give them a try.

It is not entirely clear why some teens will no longer use alcohol and drugs after an initial stage of experimentation, while others become long-term users and abusers. Experts have recently offered pro and con opinions on the existence of an addictive personality type. For some who work with addicts, there is clinical evidence of a type of personality that can best be described as impulsive and lacking moderation. Other experts believe that it is an antisocial type of person who is most susceptible to becoming a substance abuser (Thiers, 4).

CURIOSITY

The teen years involve experimentation with many kinds of activities. One teen recently said, "I won't know if it's right for me unless I try." Teens are anxious and must explore. This sense of wanting to know about things firsthand is pervasive, and extends to experimentation with mood-altering chemicals. When one enjoys peak health and prospects of longevity, there is a sense of immortality approaching a subconscious belief in magic. The magic belief is "I am young and energetic with all of life ahead of me, and no harm can come to my body!"

ALIENATION AND EMPTINESS

Young peeple who are heavy users of drugs and/or alcohol often tell us that they are turned off by families and schools. They sometimes perceive themselves as alone or outside of the mainstream of life. Some say that society is "screwed up" and hopeless. Getting high is a way to feel okay in the midst of a world that has gone sour. Drugs and alcohol enable one to escape from a hostile and uncaring environment. The future—if it comes up for consideration at all—appears empty and without promise to these users. This is the bleak outlook of the alienated (Jessor and Jessor; Glassner and Loughlin; Carpenter, Glassner, Johnson, and Loughlin).

OTHER REASONS

Some deal in drugs for the "big bucks," and then become users. Others are users who become dealers to support a habit. Ease of availability of drugs and alcohol are also cited as reasons for getting into them. Some express curiosity by trying drugs just because they're out there. Much remains to be learned about this entire area of why some become users/abusers, and others do not (Glassner and Loughlin, 41–64).

REFERENCES

Barnes, Deborah. "Drugs: Running the Numbers," *Science* 240, no. 4859 (June 1988): 1729–1731.

Benson, Peter L., Dorothy Williams, and Arthur L. Johnson. *The Quicksilver Years*. San Francisco: Harper & Row, 1987.

Bridgers, Sue Ellen. *Permanent Connections*. New York: Harper & Row, 1987.

"Campus Dryout," *Time* 131, no. 23 (6 June 1988): 70.

Carpenter, Cheryl, Barry Glassner, Bruce Johnson, and Julia Loughlin. *Kids, Drugs, and Crime*. Lexington, MA: Lexington Books, 1988.

Donovan, Mary Ellen. "Alcohol Abuse: What Your Child Should Know," *Parents* 63, no. 2 (February 1988): 189.

Estes, Nada, and M. Edith Heinemann. *Alcoholism*. 2d ed. St. Louis, MO: C. V. Mosby, 1982.

"Getting Tough," *Newsweek* 108, no. 2 (22 September 1986): 40.

Gibbs, Mary. "When Parents Just Say No," *Time* 131, no. 7 (15 February 1989): 94.

Glassner, Barry, and Julia Loughlin. *Drugs in Adolescent Worlds*. New York: St. Martin's, 1987.

Groller, Ingrid. "Drugs and Alcohol: Which Is Worse?" *Parents* 63, no. 1 (January 1988): 26.

Hazelden Foundation. *Learn about Cocaine*. Center City, MN: Hazelden Educational Materials, 1984.

Hundley, Tom. "Detroit Drug Dealers Spreading Out," *Chicago Tribune*, 11 December 1988: I-6.

Jessor, Richard, and Shirley Jessor. *Problem Behavior and Psychosocial Development*. New York: Academic Press, 1977.

Johnston, Lloyd D., Patrick M. O'Malley, and Jerald Bachman. *Illicit Drug Use, Smoking and Drinking by America's High School Students, College Students and Young Adults*. Rockville, MD: U.S. Dept. of Health and Human Services (NIDA), 1988.

"Kids and Cocaine," *Newsweek* 107, no. 11 (17 March 1986): 58–61.

Kinney, Jean, and Gwen Leaton. *Loosening the Grip—A Handbook of Alcohol Information*. 2d ed. St. Louis, MO: C. V. Mosby, 1983.

Lamar, Jacob. "Kids Who Sell Crack," *Time* 131, no. 19 (9 May 1988: 20–26.

"National High School Senior Survey," University of Michigan, Ann Arbor, Institute for Social Research. Press release, 13 February 1990.

Plummer, William. "Bruce Kimball Dove to Fame But Drove to Tragedy," *People Weekly* 30, no. 8 (22 August 1988): 87–88.

Ray, Oakley. *Drugs, Society, and Human Behavior*. 3d ed. St. Louis, MO: C. V. Mosby, 1983.

"Teen Drug Use Continues Decline According to U.M. Survey," University of Michigan News and Information Services. Press release, 29 February 1989.

Thiers, Naomi. "The Addictive Personality," *Guidepost* 31, no. 6 (22 June 1989): 1–4, 11.

Wolff, Virginia Euwer. *Probably Still Nick Swansen*. New York: Henry Holt, 1988.

Resources
for Finding Out about Substance Abuse

Fiction

Butterworth, W. E. **Under the Influence.** New York: Four Winds, 1979. 93p.

Keith Stevens is a teenage alcoholic who admits his problem, but can't quite get around to doing anything about it. He gets into fights, is involved in accidents with his car, and is generally "a mess."

Cheatham, K. Follis. **The Best Way Out.** New York: Harcourt Brace Jovanovich, 1982. 168p.

Fourteen-year-old Haywood thinks drinking can help him escape from the problems in his life, but it is an innovative school program that shows him how best to use his talents.

Cole, Barbara. **Alex the Great.** New York: Rosen Publishing, 1989. 200p.

Concepts of loyalty and responsibility are tested when Deonna, a high school tennis star, must decide what to do about her best friend Alex's drug problem.

Due, Linnea. **High and Outside.** New York: Bantam, 1980. 195p.

Niki's parents have always allowed and even encouraged her drinking at home, not recognizing how she uses liquor to ease the pressure of her life as a good athlete and student. When her status is jeopardized, she turns to AA.

Greene, Shep. **The Boy Who Drank Too Much.** New York: Dell, 1979. 157p.

Buff, a good athlete, feels like an outsider after his mother's death. Because of abuse and pressure from his father, he turns to alcohol. His friends help him to see what's happening, and to separate from his father.

Horwitz, Joshua. **Only Birds and Angels Fly.** New York: Harper & Row, 1985. 186p.

Danny enjoys making friends with Chris Jordan while he is at Connel School. Only much later does he learn about Chris's involvement in drugs and the fast life. Taking off with Chris on a wild fling to Colorado seems okay, until Danny is forced to reexamine his own basic values.

Kropp, Paul. **Dope Deal.** St. Paul, MN: EMC Publishing, 1979. 93p.

Brian has earned a living pushing marijuana, but when he returns to his father's house after his arrest and discovers that his younger brother is involved in drugs, he changes his values.

———. **Snow Ghost.** St. Paul, MN: EMC Publishing, 1984. 93p.

Martin suffers from low self-esteem, and is the product of a depressed upbringing. He tries to solve his problems with drugs and antisocial behavior, but a wilderness experience with a caring counselor changes his life.

Levy, Marilyn. **Summer Snow.** New York: Ballantine, 1986. 171p.

Leslie is spending the summer with her father and his girlfriend when she gets in with a fast crowd using drugs. Before she is totally out of control, she gets help from a reformed addict.

Norris, Gunilla B. **Take My Walking Slow.** Atheneum, 1970. 89p.

This is a story about Richie and his alcoholic father. Richie's family is split apart as a result of alcoholism, but he manages in the long run to deal with his problems.

Roos, Steven. **You'll Miss Me When I'm Gone.** New York: Delacorte, 1988. 157p.

Marcus starts sneaking drinks at home to help himself adjust to his parents' divorce and the pressure of trying to keep up at school. When his drinking increases, he risks losing his girlfriend and his position at school unless he admits to himself what he is.

Strasser, Todd. **Angel Dust Blues.** New York: Coward, McCann & Geoghegan, 1979. 203p.

Alex's involvement with Michael leads him to a "job" dealing drugs. Alex also develops a relationship with Ellen, and would soon like to quit dealing drugs—especially after seeing what it does to Michael's life.

Nonfiction

Bartimole, John, and Carmella Bartimole. **Teenage Alcoholism and Substance Abuse: Causes, Cures and Consequences.** Hollywood, FL: Compact Books, 1986. 160p.

This book examines some of the social backdrop of teenage substance abuse. Valuable information is provided in a clear and straightforward manner.

Barun, Ken, and Philip Bashe. **How To Keep the Children You Love off Drugs.** New York: Atlantic Monthly Press, 1988. 288p.

A presentation of the facts concerning drugs and drug abuse. A valuable feature is a list of drug treatment programs in all states, including where to write for more information on substance abuse.

Berger, Gilda, and Melvin Berger. **Drug Abuse—A–Z.** Hillside, NJ: Enslow, 1990. 144p.

This is a reference on drugs and drug-related terminology, including slang terms. In addition to drug information, the authors provide a history of governmental regulation of illegal substances.

Brook, Judith S., Dan Lettieri, David W. Brook, and Barry Stimmel. **Alcohol and Substance Abuse in Adolescence.** New York: Haworth Press, 1985. 216p.

An excellent reference book for all who need a deeper understanding of adolescent drug use. The physical as well as social aspects of drug use are explored.

Cohen, Daniel, and Susan Cohen. **A Six-Pack and a Fake ID.** New York: M. Evans, 1987. 156p.

Lots of facts and straight information about alcohol abuse. The chapter titled "You Can Vote, but You Can't Drink" is especially forceful.

Coombs, Robert H., ed. **The Family Context of Adolescent Drug Use.** New York: Haworth Press, 1988. 265p.

Adolescent drug use is examined from the point of view of family systems and differing styles of parenting. This volume offers deep insights into the relationship between family dynamics and control on one hand, and teen drug use on the other.

Dolmetsch, Paul, and Gail Mauricetti, eds. **Teens Talk about Alcohol and Alcoholism.** Garden City, NY: Doubleday, 1987. 144p.

This handbook was written in large part by the students at Mount Anthony Union Junior High School in Vermont. It is easy to read and covers such topics as living with an alcoholic and recognizing alcohol abuse.

Edwards, Gabrielle I. **Coping with Drug Abuse.** New York: Rosen Publishing, 1990. 150p.

This book details the history, myths, and perils of marijuana, hallucinogens, cocaine, heroin, tobacco, and many other substances. This revised edition has new information on crack.

Grosshandler, Janet. **Coping with Alcohol Abuse.** New York: Rosen Publishing, 1990. 150p.

This book carefully examines teenage alcohol abuse, including: family and peer environments that foster alcohol abuse, coping skills to prevent abuse, and a guide through counseling and recovery from alcoholism for teens.

———. **Coping with Drinking and Driving.** New York: Rosen Publishing, 1990. 150p.

This concise book offers statistics about the effects of drinking and driving, facts about what alcohol does to the body, and possible solutions to the problem, such as pacts with parents, separating driving from drinking, and peer cooperation. Contains information on where to go for help, including such organizations as SADD and MADD.

Hawley, Richard A. **Drugs and Society—Responding to an Epidemic.** New York: Walker & Co., 1988. 157p.

This is an expert discussion of the kinds of drugs that are commonly used, how they are used, and the effects they produce. At the end of each chapter there are key questions for review of the material. Extensive appendices offer excellent reference material.

Kinney, Jean, and Gwen Leaton. **Loosening the Grip—A Handbook of Alcohol Information.** 3d ed. St. Louis: C. V. Mosby, 1986. 369p.

This is one of the best informational books dealing with alcohol and its effects. Chapter 10 deals with special populations, including adolescents. Chapter 8, "Treatment," is one of the very best on this topic.

Leite, Evelyn, and Pamela Espeland. **Different Like Me.** Minneapolis, MN: Johnson Institute Books, 1987. 110p.

Aimed at teens who worry about the substance abuse of their parents. Factual but brief attention is given to symptoms, long-term consequences, and what a teenager can do to help.

McFarland, Rhoda. **Coping with Substance Abuse.** New York: Rosen Publishing, 1987. 145p.

An account of drug and alcohol abuse. This book deals at length with family dynamics of chemically dependent people, including codependency.

Newman, Susan. **You Can Say No to a Drink or a Drug.** New York: Perigee, Putnam, 1986. 171p.

Each chapter tells a story about teens—those who use drugs and alcohol, and those who abstain. The dialogue is realistic and avoids a preachy type of moralizing. Each chapter ends with facts and statistics under the heading "You Should Know."

Rhodes, Jean E., and Leonard Jason. **Preventing Substance Abuse among Children and Adolescents.** Elmsford, NY: Pergamon Press, 1988. 151p.

Some current research findings and an extensive discussion of prevention programs are the focus of this excellent book. There are some concrete ideas offered for prevention programs of the future.

Strack, Jay. **Drugs and Drinking.** Nashville: Thomas Nelson Publishers, 1985. 228p.

Examines motives for substance abuse, as well as suggestions for parents, churches, schools, and the larger society. The chapter entitled "Cocaine: The Feel-Good Drug" is especially well done and pertinent.

Tessler, Diane. **Drugs, Kids, and Schools.** Santa Monica, CA: Goodyear, 1980. 210p.

This book offers much factual information, with long lists of reference materials at the end of each chapter, plus a huge amount of appendix material.

U.S. Department of Education. **Schools without Drugs,** Washington, DC: U.S. Government Printing Office, 1987. 78p.

Provides summary information about drugs and drug education. Includes sources of information, school and community resources, and hotlines.

Nonprint Materials

Are You Talking to Me?

Type: 16mm film, video; color

Length: 29 min.
Cost: Rental $81 (film or video); purchase $595 (film), $495 (video)
Distributor: Kinetic, Inc.
255 Delaware Avenue, Suite 340
Buffalo, NY 14202
Date: 1989

Teenagers in this film tell about the effects of drugs on their lives. The many problems associated with substance abuse are featured.

Back from Drugs
Type: Various video; color
Length: 30 min.
Cost: Rental $95 (seven days, including shipping time) (1/2″ VHS only); purchase $150 (technically a license)
Distributor: PBS Video
1320 Braddock
Alexandria, VA 22314-1698
(800) 424-7963
Date: 1986

The teens in this film explain how they became drug users, and how they were motivated to seek rehabilitation.

Choices and Consequences
Type: 16mm film, various video; color
Length: 30 min.
Cost: Rental $50 (one week)(free preview); purchase $525 (film), $495 (video)
Distributor: FMS Productions
P.O. Box 4428
520 E. Montecito St., Suite F
Santa Barbara, CA 93140
(800) 421-4609 (outside CA)
(800) 564-2488 (within CA)
Date: 1987

Tells the story of three teenagers who become involved with alcohol and drugs. The emphasis is on the difficulties in getting teens to face up to substance abuse.

Cocaine and Crack—The Lost Years

Type: Various video; color
Length: 23 min.
Cost: Rental $75 (three days)(plus shipping charges); purchase $560 (film), $300 (video)
Distributor: Encyclopedia Brittanica Films
425 N. Michigan Ave.
Chicago, IL 60611
(800) 554-9862
Date: 1987

Teens who are lonely and trying to establish an identity are shown experimenting with drugs. Also featured are teenagers in a drug rehab program describing how they became abusers.

Cocaine—Athletes Speak Out

Type: Various video; color
Length: 34 min.
Cost: Rental $75 (three days) (plus shipping charges); purchase $189
Distributor: Encyclopedia Brittanica Films
425 N. Michigan Avenue
Chicago, IL 60611
(800) 554-9862
Date: 1987

Interviews with professional athletes, coaches, and former drug users in this film show clearly how coke takes control of the lives of users.

Cruel Spirits: Alcohol and Violence

Type: 16mm film, video; color
Length: 32 min.
Cost: Rental $75 (film or video); purchase $595 (film), $495 (video)
Distributor: Coronet/MTI Film & Video
108 Wilmot Road
Deerfield, IL 60015
(800) 621-2131; (708) 940-3600
(call collect from Illinois and Alaska)
Date: 1989

This film presents up-to-date information on the physical and social effects of drinking. Among its many positive aspects is its explanation of the role which denial plays in the judgment of problem drinkers.

Diagnosis: Teen Alcoholic

Type: Nonmotion video; color
Length: 13 min.
Cost: $99
Distributor: Encyclopedia Brittanica Educational Corp.
310 S. Michigan Avenue
Chicago, IL 60604
(800) 554-9862
Date: 1986

Ginny, a teenager, overcomes her alcoholism in this program, which presents slides in a videotape format. Information is given on how alcohol dependency develops and what teens can do to avoid it.

Drinking and Driving—The Toll, the Tears

Type: Various video; color
Length: 60 min.
Cost: Purchase $44.95
Distributor: WETA
Box 2626
Washington, DC 20013
(800) 445-1964
Date: 1986

This documentary film, introduced by Phil Donahue and originally shown on PBS, features interviews with victims and families as well as drunk drivers themselves. It comes with a discussion guide.

Drugs, Alcohol and Tobacco

Type: Sound filmstrip or video series; color
Length: 12 min. each (filmstrip or video)
Cost: Purchase $116 (three filmstrips), $124 (three videos)

Distributor: Encyclopedia Brittanica Educational Corp.
310 S. Michigan Avenue
Chicago, IL 60604
(800) 554-9862
Date: 1987

These programs show the effects of these substances on the body and on behavior. A special feature is the emphasis on resisting peer pressure and establishing drug-free alternatives. This series is also available in Spanish.

Keeping Kids off Drugs

Type: Various video; color
Length: 25 min.
Cost: Rental $75 (one day)(plus $5.85 shipping); purchase $149 (VHS or Beta); other prices on request
Distributor: Films For the Humanities and Sciences
P.O. Box 2053
Princeton, NJ 08543
(800) 257-5126
Date: 1988

Schools, families, and community agencies are instructed on what they can do to prevent at-risk youth from becoming substance abusers.

Kids and Drugs

Type: VHS or Beta; color
Length: 28 min.
Cost: Rental $75 (one day)(plus $5.75 shipping); purchase $149
Distributor: Films for the Humanities and Sciences
P.O. Box 2053
Princeton, NJ 08543
(800) 257-5126
Date: 1987

A story of how five teenagers struggle with drugs, including cocaine and heroin. Of special interest is the denial that users indulge in at the early stages of drug abuse.

Kids Talk to Kids about Drugs

Type: Various video; color
Length: 28 min.
Cost: Rental $75 (one day)(plus $5.75 shipping); purchase $249
Distributor: Films for the Humanities and Sciences
P.O. Box 2053
Princeton, NJ 08543
(800) 257-5126
Date: 1988

A group of New York City teens tell about the peer pressures and hopelessness that led them into drugs, and how they finally learned to say no.

Medical Effects of Alcohol

Type: Various video; color
Length: 13 min.
Cost: Rental $75 (three days)(plus shipping charges); purchase $325 (film), $230 (video)
Distributor: Encyclopedia Brittanica Films
425 N. Michigan Avenue
Chicago, IL 60611
(800) 554-9862
Date: 1984

A feature of this film is the description of damage to body organs from habitual social drinking. A no-nonsense approach is taken to the knowledge that alcohol is a drug and that physical damage to a drinker's body is almost inevitable.

New Crack Facts

Type: Video; color
Length: 14 min.
Cost: Purchase $295
Distributor: Perennial Education, Inc.
930 Pitner Avenue
Evanston, IL 60202
(708) 328-6700
Date: 1989

Factual information on crack cocaine is presented, partially through two interviews with teenage users. The video explains why crack is so addictive to most users.

Second Half

Type: 1/2" VHS or 3/4"; color
Length: 29 min.
Cost: Rental (call for free preview); Purchase $385
Distributor: FMS Productions
P.O. Box 4428
520 E. Montecito St., Suite F
Santa Barbara, CA 93140
(800) 421-4609 (outside CA)
(805) 564-2488 (within CA)
Date: 1990

The story of Thomas "Hollywood" Henderson, whose pro football career was lost to cocaine. Focuses on how he overcame addiction in 1983 and has kept clean since.

Sex and Drugs: The Intimate Connection

Type: VHS and U-Matic; color
Length: 30 min.
Cost: Rental (call for free preview); purchase $385
Distributor: FMS Productions
P.O. Box 4428
520 E. Montecito St., Suite F
Santa Barbara, CA 93140
(800) 421-4609 (outside CA)
(800) 564-2488 (within CA)
Date: 1990

Three recovering addicts are the focus of this film, which explores the connection between drugs and intimacy. Appropriate for teens as well as adults.

Smoking—Hazardous to Your Health

Type: Various video; color
Length: 29 min.
Cost: Rental $95 (seven days, including shipping time, 1/2" only); purchase $200 (technically a license)

Distributor: PBS Video
1320 Braddock
Alexandria, VA 22314-1698
(800) 424-7963
Date: 1987

This is the first of a two-part series about the effects of smoking upon the body. Teens will be interested in the vivid description of what smoking does to hearts and lungs.

Teen Addiction
Type: VHS or Beta; color
Length: 19 min.
Cost: Rental $75 (one day)(plus $5.75 shipping); purchase $149
Distributor: Films for the Humanities and Sciences
P.O. Box 2053
Princeton, NJ 08543
(800) 257-5126
Date: 1987

This is the account of an 18-year-old's struggle with drugs and alcohol. His parents tell about the effects of his substance abuse upon the family. Finally, he gets into a treatment program and overcomes his addiction.

Teenage Alcoholism
Type: Various video; color
Length: 24 min.
Rent from: Boston University
Krasker Memorial Film Library
565 Commonwealth Avenue
Boston, MA 02215
(617) 353-3272 (for cost and arrangements)
Date: 1982

This film focuses on the increasing problem of alcoholism among teenagers. A 16-year-old has become alcoholic, and viewers can see his struggle with the addiction and the difficulty his parents experience in attempting to accept the fact.

Teenage Drinking and Drug Abuse
Type: Various video; color

Length: 30 min.
Cost: Rental $95 (seven days, including shipping time)(1/2″ only); purchase $150 (technically a license)
Distributor: PBS Video
1320 Braddock
Alexandria, VA 22314-1698
(800) 424-7963
Date: 1986

Listen to what teenagers themselves say about their use of drugs and alcohol. This film focuses on peer pressure and the belief of some that they can use substances without becoming abusers.

Teenage Substance Abuse: An Open Forum with John Callahan
Type: 16mm film, video; color
Length: 23 min.
Cost: Rental $75; purchase $495 (film), $325 (video)
Distributor: Coronet/MTI Film & Video
108 Wilmot Road
Deerfield, IL 60015
(800) 621-2131; (708) 940-3600
(call collect from Illinois and Alaska)
Date: 1990

Based on Callahan's extensive interviews with teens, the film deals with root causes of abuse. The teens in this film tell of the denial and self-pity they experienced, as well as the pain of chemical abuse.

A Teenager's Underground Guide to Understanding Parents
Type: 16mm film, video; color
Length: 26 min.
Cost: Purchase $520 (film), $250 (video)
Distributor: Coronet/MTI Film & Video
108 Wilmot Road
Deerfield, IL 60015
(800) 621-2131; (708) 940-3600
(call collect from Illinois and Alaska)

Rent from: University of Minnesota
University Film & Video
1313 Fifth Street SE
Minneapolis, MN 55414
(800) 847-8251 (for cost and arrangements)
Date: 1983

This is an unusual approach to improving understanding and communication between teens and parents. A major focus is on drugs and alcohol, and the film points up the differing attitudes toward them of teenagers and their parents.

Too Good To Waste
Type: 16mm film, various video
Length: 20 min.
Cost: Rental $50 (one week); purchase $450
Distributor: FMS Productions
P.O. Box 4428
520 E. Montecito, Suite F
Santa Barbara, CA 93140
(800) 421-4609
Date: 1985

An awareness film, with peer-to-peer dialogue about drugs and alcohol.

Tuned Out
Type: Nonmotion video
Length: 13 min.
Cost: Purchase $99
Distributor: Encyclopedia Brittanica Educational Corp.
310 S. Michigan Avenue
Chicago, IL 60604
(800) 554-9862
Date: 1983

This program, using slides in a film format, tells the story of Louis, who was "tuned out" because of drug abuse. Viewers learn the effects of many of the popular "mind-bending" drugs, including cocaine and speed.

You Can Say No to a Drink or a Drug
Type: 16mm film, video; color

Length: 30 min.
Cost: Purchase $550 (film), $495 (video)
Distributor: Coronet/MTI Film & Video
108 Wilmot Road
Deerfield, IL 60015
(800) 621-2131; (708) 940-3600
(call collect from Illinois and Alaska)
Date: 1988

Shows several situations in which teens must make decisions about using drugs and alcohol.

Young People and Alcohol

Type: VHS, U-Matic; color
Length: 15 min.
Cost: Rental (call distributor for free preview); purchase $195
Distributor: FMS Productions
P.O. Box 4428
520 E. Montecito Street, Suite F
Santa Barbara, CA 93140
(800) 421-4609 (outside CA)
(800) 564-2488 (within CA)
Date: 1989

Teenagers are helped to make informed choices about alcohol use by this film, which also provides information about the effects of alcohol on the user's body.

Young People and Drug Abuse

Type: VHS, U-Matic; color
Length: 15 min.
Cost: Rental (call distributor for a free preview); purchase $195
Distributor: FMS Productions
P.O. Box 4428
520 E. Montecito Street, Suite F
Santa Barbara, CA 93140
(800) 421-4609 (outside CA)
(800) 564-2488 (within CA)
Date: 1989

The reasons young people use drugs and the risks of drug abuse are fully explored in this video.

Organizations

Al-Anon and Alateen
P.O. Box 862, Midtown Station
New York, NY 10018
(212) 302-7240
Public Information Officer: Carol Kuney

Provides support to families of alcoholics. Alateen is for youth, ages 12 to 20 years, whose lives are affected by living with an alcohol abuser.

PUBLICATIONS: *Alateen Talk,* bimonthly.

Help, Inc.
638 South Street
Philadelphia, PA 19147
(215) 546-7766
Executive Director: William H. Bruce

Provides drug education for all ages. Maintains a telephone counseling service and referrals for drug users.

Drugs Anonymous
P.O. Box 473, Ansonia Station
New York, NY 10023
(212) 874-0700
Secretary: Mary Lou Phippen

Provides assistance to teens and adults who abuse drugs; including tranquilizers, sedatives, cocaine, and marijuana.

Students against Driving Drunk (SADD)
P.O. Box 800
Marlboro, MA 01752
Executive Director: Robert Anastas

Urges students to avoid drunk driving through awareness programs and speakers.

PUBLICATION: *SADD Update,* quarterly.

Hotlines

Alcohol and Drug Helpline

(800) 821-4357

The Helpline offers referral to local treatment centers and support groups.

Cocaine Abuse Hotline

(800) 992-9239

This hotline provides crisis intervention and information about treatment for abuse of all substances, not just cocaine.

NIDA Hotline

(800) 662-HELP

NIDA (National Institute on Drug Abuse) gives information about drugs and directs callers to local treatment facilities.

CHAPTER 4

Teen Runaways

With Ruth K. J. Cline

> Willie didn't leave a note. If his parents knew he was actually leaving, they'd check the bus station and Amtrak and if they found out which way he headed, they could have him picked up in Spokane. . . . No way. He'll write later; a postcard from Spokane maybe, saying he's headed for Seattle. Gotta do this right. Gotta disappear right; take appropriate evasive action. First thing he's done right since the accident.
>
> Chris Crutcher, *The Crazy Horse Electric Game* (New York: Greenwillow, 1987), 89.

Willie can't think clearly about his options because he is too emotionally involved in the decisions. His parents, his girlfriend, and his friends are all trying to help him, but Willie can't see any way to deal with his problems except to leave home.

Historical and Romanticized Runaways

In the past, running away from home was considered a natural part of growing up. After a child was reprimanded, he or she might pack a few things in a bag and take off for grandma's, or tromp around the block. It may have been a short distance or a short time span, but it was a release for hurt feelings. There was no harm done, and forgiveness came from both sides.

The Adventures of Huckleberry Finn, by Mark Twain, is considered a classic runaway story: Huck and his friend Jim raft down the Mississippi River. Behind the humor and innocence of this story is the serious abuse Huck received from his father, forcing Huck to run away to save his life. The river trip, which many readers think of as idyllic, was fraught with dangers, voices in the night, and fear in the hearts of the two characters.

Few who run away today fall into the Huck Finn model. "For some it is simply a matter of escaping from unbearable, humiliating physical punishment or sexual abuse. For many more, running away feels like a desperate assertion of selfhood. Many young people no longer can be or wish to be the good child their parents seem to insist on" (Gordon, 9).

Young people have always yearned for the quest for fortune, the journey around the world, and the search for Prince Charming. In their minds, they have idealized leaving home and being independent. Young people need to realize that the environment today is full of hazards that weren't even suspected in earlier times.

Statistics and Facts about Runaways

- Approximately two million children and adolescents disappear each year in the United States. Of these, 1,850,000 are runaways, 100,000 are taken by a parent, and 50,000 simply disappear (Janus, 10).
- Runaways can't work legally without work permits, which parents or guardians must approve (Brenton, 42).
- Doctors usually require a parent or guardian's approval before they will treat a minor (ibid.).
- Schools will usually not allow a minor to register without a parent or guardian's approval (ibid.).
- Running away from home is a *status crime* in the United States, and teens can be arrested if they are unable to prove they are living with a parent or guardian (Raphael and Wolf, 6).

- The majority (80–90 percent) of runaways are back home within 48 hours (Langway, 98).
- Approximately 60 percent of runaways are girls (ibid., 97).
- About 36 percent run away because they were physically or sexually abused at home (Hersch, 31).

Definitions

Runaways were once thought of as capricious and undisciplined nomads. Today they represent a young population, hurt and looking for freedom. They are often family oriented, between the ages of 9 and 16, and referred to authorities by themselves or school (Janus, 11). Many of them face life-threatening dangers as they become involved in prostitution, drugs, or other street crime for survival.

"A **throwaway** is a child who is most often thrown from his home or asked to leave by a very angry, a very depressed, a very drunk parent, [or] a parent high on drugs who simply cannot cope with his or her own problem. These children are forced out. It is very rare that these kids want to run. Most children want to go home. They are not carefree children" (ibid., 165). Some youths are labeled "independent oriented" with either no family or a family that no longer offers a viable living arrangement (U.S. House of Representatives, 162).

Missing children are less than 18 years old and are the innocent targets of criminal activity. Sometimes they are abducted from the parent who has legal custody (U.S. House of Representatives, 35).

Status offenders have committed no criminal act, but are confined in secure facilities (U.S. House of Representatives, 45). As a group, they are less likely to repeat the offense, their offense careers are different from the throwaway type, and the seriousness of this label can be aggravated by legal processing of the individual (U.S. House of Representatives, 51).

Causes of Running Away

There is no segment of the population that is free from the specter of the child/adolescent runaway: economic conditions and social background of the family are almost irrelevant. An executive's young daughter who is tired of the curfews and restrictions on her dating activities is as likely to run away as the son of a blue-collar worker who is angry because his parents nag him about drugs. It cannot be said that one kind of parenting will automatically trigger runaway behavior, but parents need to be aware of their children's activities and communicate with them in a reasonable fashion. Also, it should be noted that in one study of teenage runaway girls in the San Francisco area, 60 percent reported that they had been previously abused by at least two males. One-third of them had been abused by their natural fathers (Silbert and Pines, 285).

Parental inflexibility may be caused by the insecurity parents feel in dealing with their teenager (Brenton, 88). Parents may think it is a sign of weakness if they do not act in a tough manner with their children. Rigidity in parental control, however, can be a strain on some adolescents, and they need to get away from it. Ineffective communication between parent and teenager is a frequently mentioned problem (Palenski and Launer, 350–351). Of course school failures are often a contributing factor. Parents and children often fight about schoolwork, becoming locked in a power struggle in which compromises are ineffective.

There are also teen runaways from families where permissiveness is practiced. Children from such homes may feel their parents don't care about them because they don't set limits. Everything is accepted. Some teens also expressed concern over parents who never showed affection towards them: no passion, no arguing, no involvement (ibid., 91).

Brenton concludes that teenagers need guidelines that they have helped to establish. They need structure and rules that are applied with a reasonable degree of consistency. They need a sense of trust and self-respect (Brenton, 23–48).

In *Adolescent Runaways,* quite a different picture is painted of the family influence on the runaway (Janus, 5–62). The authors based their conclusions on a study of 149 runaways who came to Covenant House in Toronto, Canada. Because the study has a small number of participants, all self-reported runaways, it may not be possible to generalize to a broad segment of the population. However, the results are startling and provide the basis for concern about our society.

The runaways indicated they first left home from as early as 4 years old to as late as 19 years old. The study, including 63 percent males and 37 percent females, reported leaving home frequency to be from 1 to 110 times, with 8 times representing the average. A large proportion of the respondents reported physical and/or sexual abuse at home, with 73 percent reporting they had been physically beaten. Forty-three percent stated that being physically abused by the people they lived with was an important reason for their leaving home. Fifty-one percent of the runaways reported sexual abuse.

Consequences of Running Away

RETURNING OR BEING RETURNED HOME

Chances are that the runaway will stay close to familiar surroundings and friends, and be home within a few days. Forty percent of all runaways are back home the very next day, 60 percent within three days. Half travel fewer than 10 miles from home, with only one in five runaways venturing farther than 50 miles. Sixty percent stay with friends (Lonnborg and McCall, 182).

If the teen is gone overnight or longer, parents may decide to call the police. Lonnborg and McCall discourage this idea because "police are responsible for returning only eighteen percent of all runaways" (182). They suggest police do not have the personnel to carry on intensive searches for the many reported missing children. Police have a limited number of options open to them, and may not be required by law to return the child to the family. The treatment of status offenders varies from state to state. In some states, the runaway is placed in a jail with adult offenders or in a juvenile detention center with delinquents. In other states, the runaway may not legally be held in a secure facility for longer than overnight or up to forty-eight hours (Lonnborg and McCall, 184).

The family can carry on their own search for the teenager by calling friends or relatives. Parents can drive by an area or hang-out that the adolescent may frequent. A call to the school might bring results, as it did for one family who discovered their runaway daughter was attending classes regularly (Lonnborg and McCall, 184).

PROSTITUTION

Regrettably, a significant number of runaways do not return home soon, if at all. They face a bleak future with considerable danger. Being homeless and without funds in cities where they don't have family or friends means barely surviving on the streets. They become part of an estimated 300,000 children and teens who are hard-core street kids (Hersch, 31). Here they are usually forced into prostitution or other street crime. They aren't just risking a reputation for being bad kids by selling sex. They are often at the mercy of pimps and perverts who view them as a disposable commodity. "Street kids die quickly and quietly. More than 5,000 teenagers a year are buried in unmarked graves" (Axthelm, 64).

It is estimated that there are over a million teen runaways on the streets of this country, and a significant number of them are hustling sex and drugs for survival (Hersch, 31). In addition to the physical dangers of the street life, ranging from injury to infection and AIDS, a large number of these runaway prostitutes were judged to be emotionally damaged (ibid., 287). Research suggests that poor self-esteem and lack of sex education characterize these teens (Schaffer and DeBlassie, 691–692).

The connection between runaways and prostitution is not limited to girls. Boys, though not necessarily homosexuals themselves, often fall into the same trap as girls. They are known on the streets as "chickens" and those who exploit them sexually are "chicken hawks." One researcher of the New

York street scene described the pathetic plight of a couple, both teenage runaways:

> William, 19, and Susan, 17, walk by at 1 a.m. pushing their 6-month-old child in a stroller among the drug dealers, hustlers, and addicts. Susan continues to walk with the baby as William hustles nearby, peddling sex to male customers. William is not gay, but he needs money to buy diapers and food for his baby. Perhaps later he will make love to Susan.
>
> (Hersch, 30)

The Home Environment

Since the families of runaways are often dysfunctional, the runaway usually should not be returned to that home environment, unless changes occur. The Janus study substantiated findings of greater verbal abuse and alcohol and drug misuse in the families of physically abused children (Janus, 36). Family therapy in which all members of the family participate in attempting to improve communication as well as the emotional climate within the home is often needed. Where physical or sexual abuse is involved, intervention by governmental social agencies is necessary.

When parents realize something has gone wrong to cause the running away in the first place, they should try to correct the situation. Parents often have feelings of guilt about the runaway and pass those feelings on to the child. Some parents are quick to get professional help for their child without thinking about themselves and what is lacking in their part of the relationship (Raphael, 169).

Some parents have been able to work out compromises with their children so that the act of running away proves to open a door to a better relationship (Raphael, 171). Conflict may be an inevitable part of all social interaction, but at what point does it become destructive? There is a continuum for abuse, whether sexual or physical, and the participants of such activity need to recognize where they are on this continuum and when to get help.

REFERENCES

Axthelm, Pete. "Somebody Else's Kids," *Newsweek* 111, no. 17 (25 April 1988): 64–68.

Berman, Claire. "The Runaway Crisis," *McCall's* 115, no. 4 (January 1988): 113–116.

Brenton, Myron. *The Runaways: Children, Husbands, Wives and Parents*. Boston: Little, Brown, 1978.

Crutcher, Chris. *The Crazy Horse Electric Game*. New York: Greenwillow, 1987.

Gordon, James S., ed. *Reaching Troubled Youth: Runaways and Community Mental Health*. Washington, DC: U.S. Department of Health and Human Services, 1981.

Hersch, Patricia. "Coming of Age on City Streets," *Psychology Today* 22, no. 1 (January 1988): 28–37.

Janus, Mark-David, Arlene McCormack, Ann Wolbert Burgess, and Carol Hartman. *Adolescent Runaways: Causes and Consequences*. Lexington, MA: Heath, 1987.

Langway, L. "A Nation of Runaway Kids," *Newsweek* 100, no. 16 (18 October 1982): 97–98.

Lonnborg, Barbara, and Robert B. McCall. "Unlikely Runaways," *Parents* 59, no. 12 (December 1984): 180–184.

Palenski, Joseph E., and Harold Launer. "The Process of Running Away: A Redefinition," *Adolescence* 22, no. 86 (Summer 1987): 347–362.

Raphael, Maryanne, and Jenifer Wolf. *Runaways: America's Lost Youth*. New York: Drake, 1974.

Schaffer, Bernie, and Richard DeBlassie. "Adolescent Prostitution," *Adolescence* 19, no. 75 (Fall 1984): 689–696.

Silbert, Mimi H., and Ayala Pines. "Early Sexual Exploitation as an Influence in Prostitution," *Social Work* 28, no. 4 (July–August 1983): 285–289.

Twain, Mark. *The Adventures of Huckleberry Finn*. Boston: Houghton Mifflin, 1958. 245p.

U.S. House of Representatives. *Juvenile Justice, Runaway Youth, and Missing Children's Act Amendments of 1984*. Washington, DC: U.S. Government Printing Office, 1984.

Resources
for Finding Out about Teen Runaways

Fiction

Anonymous. **Go Ask Alice.** Englewood Cliffs, NJ: Prentice-Hall, 1971. 189p.

The diary of a teenager who runs away from home, gets involved in drugs, and dies of an overdose after she decides to go straight.

Arrick, Fran. **Steffie Can't Come Out To Play.** New York: Bradbury, 1978. 160p.

Fourteen-year-old Stephanie runs away from her dreary industrial Pennsylvania town to be a model in New York. Instead she meets an expensively dressed man who puts her to work for him.

Bonham, Frank. **Viva Chicano.** New York: Dutton, 1970. 179p.

Seventeen-year-old parolee Joaquin Duran, tormented by his mother's nagging and haunted by his past, decides to run away after he is accused of pushing his two-year-old brother out of a window.

Butler, Beverly Kathleen. **A Girl Called Wendy.** New York: Dodd, Mead, 1976. 211p.

Wendy, a 15-year-old Native American girl living in a mission boarding school on the reservation with her 7-year-old sister, learns that her parents have divorced and the two girls must go to Milwaukee to stay with an aunt. They run away to rejoin their mother.

Calvert, Patricia. **When Morning Comes.** Scribner's, 1989. 153p.

Cat is a 15-year-old runaway from a foster home. After trying to persuade her widowed mother that she should live with her, Cat is again rejected. This novels tell much about life on the street and how a teenager can change her life.

Christopher, Matt. **Dirt Bike Runaway.** Boston: Little, Brown, 1983. 160p.

A shy, unhappy 16-year-old with a talent for working with motorcycles runs away from his foster home and gets involved with a variety of people, both bad and good.

Corcoran, Barbara. **The Hideaway: A Novel.** New York: Atheneum, 1987. 112p.

After running away from a reform school, 15-year-old Tom hides from the police, with the help of his sister, until he can clear himself of a drunk driving charge.

Crutcher, Chris. **The Crazy Horse Electric Game.** New York: Greenwillow, 1987, 215p.

Willie, 16, leads a charmed life in Coho, Montana, admired by his peers and loved by his parents, with a career in baseball ahead of him. All this changes when Willie suffers severe physical damage in a water-skiing accident. Unable to accept himself, and seeing the changes in his parents' and friends' attitude toward him, he runs away from home.

Dodson, Susan. **Have You Seen This Girl?** New York: Four Winds, 1982. 182p.

Scenes of the runaway network in New York are vivid, as are the portrayals of the manipulative characters who exploit it. However, Kathy's perhaps-temporary rescue by the wealthy Mrs. Kent is hardly the usual pattern for runaway teenagers.

Dragonwagon, Crescent, and Paul Zindel. **To Take a Dare.** New York: Harper & Row, 1982. 249p.

A teenage runaway encounters both love and loneliness when she decides to settle in a small Arkansas town.

Elfman, Blossom. **The Butterfly Girl.** Boston: Houghton Mifflin, 1980. 146p.

In this first-person narrative, a flighty young girl at odds with her parents becomes an unwed mother and struggles to find a place for herself and her child.

Everly, Jeannette. **Drop-Out.** New York: Lippincott, 1963. 189p.

A thought-provoking account of the obstacles confronting teens who try to make it on their own before they are ready.

———. **See Dave Run: A Novel.** New York: Lippincott, 1978. 127p.

Running away from an intolerable home situation, a 15-year-old boy finds that he has nowhere to go and no one to turn to.

Ferris, Jean. **Looking for Home.** New York: Farrar, Straus & Giroux, 1989. 167p.

Daphne has been struggling to cope with an abusive father. After becoming pregnant, she finds that her only escape is to run away.

George, Jean Craighead. **Julie of the Wolves.** New York: Harper & Row, 1972, 170p.

Thirteen-year-old Julie, an Eskimo girl, is running away from her young, mentally retarded husband.

———. **My Side of the Mountain.** New York: Harper & Row, 1959. 178p.

Young Sam runs away from his New York City home to some family-owned land in the Catskills. The first-person narrative details the knowledge and experience Sam gains living in the wilderness.

Gibbons, Faye. **Mighty Close to Heaven.** New York: Morrow, 1985. 183p.

Running away from his grandparents' farm, 12-year-old Dave makes his way through the Georgia mountains to rejoin his wandering father. He finds disappointment and a new appreciation for what he has left behind.

Hallman, Ruth. **Breakaway.** Philadelphia: Westminster Press, 1981. 92p.

An overprotective mother keeps a teenage boy from accepting his hearing loss and getting on with his life. The boy and his determined girlfriend run away. A sympathetic landlady and his own will to recover help him learn the skills he needs to cope successfully with his disability.

Hamilton, Morse. **Effie's Horse.** New York: Greenwillow, 1990. 208p.

Elizabeth is 15 years old and pregnant by her ex-stepfather. Her mother is unaccepting, so running away seems to be the best alternative. Unfortunately, more pain and turmoil result.

Holman, Felice. **Slake's Limbo.** New York: Scribner's, 1974. 117p.

A 13-year-old orphan treated badly by his aunt hides out in the New York subway, surviving through the kindness of a few people and his hope for a better life.

Jones, Adrienne. **Street Family.** New York: Harper & Row, 1987. 269p.

Fifteen-year-old Chancy, a runaway who dreams of a family to satisfy her need to belong, hitchhikes to LA from a Texas home for wayward girls. She must deal with the unsavory people of the street as she struggles to survive.

Kilgore, Kathleen. **The Wolfman of Beacon Hill.** Boston: Little, Brown, 1982. 192p.

A teenage runaway and a social worker are brought closer together by their shared interest in the fate of an escaped wolf struggling for survival in the streets of Boston.

King, Buzz. **Silicon Songs.** New York: Delacorte, 1990. 164p.

Max is a 17-year-old computer buff who is homeless and feels unworthy. Life on the street is a scene of despair for him until some adults and a teenage girl convince him of his worth and ability.

King, Cynthia. **Sailing.** New York: Putnam, 1982. 192p.

After two years traveling with a hippie family, 18-year-old Paul returns home to finish school and to go sailing again with the girlfriend whose father's false accusations had caused him to run away.

Lasenby, Jack. **The Lake.** New York: Oxford University Press, 1989. 167p.

Ruth runs away from home to escape the sexual demands of her stepfather. During her flight to the safety of the country near a familiar lake, she must overcome obstacles of nature as well as man.

McGraw, Eloise Jarvis. **Hideaway.** New York: Atheneum, 1983. 228p.

When his father forgets to come for him after his mother leaves on a wedding trip with her new husband, 12-year-old Jerry runs away from both of them to his grandparents' house—only to find they don't live there any longer.

Mendonca, Susan R. **Tough Choices.** New York: Dial, 1980. 136p.

A young teenager caught in a custody battle runs away as she struggles to make the right choices in a world of conflicting loyalties.

Peterson, P. J. **The Boll Weevil Express.** New York: Delacorte, 1983. 192p.

A bored northern California farm boy and a brother and sister from an orphanage decide to run away to Idaho, but find themselves down and out in San Francisco.

Pfeffer, Susan Beth. **The Year without Michael.** New York: Bantam, 1987. 164p.

The remaining members of the Chapman family try to cope with the disappearance of 14-year-old Michael.

Reeves, Bruce Douglas. **Street Smart.** New York: Beaufort Books, 1981. 222p.

A teenage girl who has been reared in a Berkeley commune and a battered and abused neighborhood boy decide to run away to San Francisco. Neither is prepared for the dangers and temptations of street life.

Roth, Arthur J. **The Caretaker.** New York: Four Winds, 1980. 216p.

A teenage boy who works winters as a caretaker for houses in a Long Island resort area suddenly finds himself also taking care of his alcoholic father and a wealthy runaway girl.

Samuels, Gertrude. **Run, Shelley, Run!** New York: Crowell, 1974. 174p.

Shelley is abused by her mother, an alcoholic, and her stepfather. The book, written as a series of flashbacks, takes a critical look at juvenile courts and prisons and the brutality encountered there.

Stoutenburg, Adrien. **Where to Now, Blue?** New York: Four Winds, 1978. 186p.

Twelve-year-old Blueberry's attempt to run away from her poor home in rural Minnesota with a six-year-old tagalong from the orphanage ends in frustration, but as a result Blueberry becomes more sure of her plans for the future.

Sweeney, Joyce. **Center Line.** New York: Dell, 1984. 246p.

Stealing their father's car and running away from home is the five Cunnigan brothers' answer to their father's alcoholic abuse. What starts as an adventure turns into fear and turmoil.

Thompson, Paul. **The Hitchhikers.** New York: Franklin Watts, 1980. 83p. Black and white photographs by Susan Kuklin.

Two lonely teenagers meet while hitchhiking to California in search of love. Running away and hitchhiking are romanticized a bit, but the lives of these two young people are not. High interest and easy vocabulary, with credible characters.

Truss, Jan. **Jasmin.** New York: Atheneum, 1982. 196p.

In order to escape from troublesome younger siblings who interfere with her homework and may cause her to fail in school, Jasmin goes to live alone in the wilderness.

Veglahn, Nancy Crary. **Fellowship with the Seven Stars.** Nashville: Abingdon, 1981. 175p.

In this first-person narrative, a teenager looks back at the time she spent with a religious cult, revealing what drew her to join and then compelled her to leave.

Nonfiction

BOOKS

Berry, James R. **Kids on the Run: The Stories of Seven Teenage Runaways.** New York: Four Winds, 1978. 105p.

Interviews with seven teenage runaways explore their motives for leaving home, how they went about it, and their experiences and reflections.

Bock, Richard, and Abigail English. **Got Me on the Run: A Study of Runaways.** Boston: Beacon, 1973. 237p.

This book is based on the belief that adolescents have sound personal reasons for running away. Narratives are related of runaways and interaction with them at the Sanctuary in Cambridge or Project Place Runaway House in Boston. The social institutions that impact the lives of kids are examined in Part II.

Brennan, Tim. **The Incidence and Nature of Runaway Behavior: Final Report.** Boulder, CO: Behavioral Research and Evaluation Corp., 1975. 606p.

A study of runaways in a chosen locality indicates that 1.8 percent of the total youth population and 3.8 percent of households with youths had runaway episodes of serious intent, meaning a duration of 24 hours or longer. Requests for affordable family counseling services and well-advertised runaway shelters were prevalent. The behavioral data indicate five generalized models: (1) spontaneous unplanned episodes, (2) deliberate successful episodes, (3) temporary "good time" escapades, (4) difficult long-term escapist episodes, and (5) temporary escapist episodes from unpleasant home situations, where intent was to stay away for only a few days.

Brenton, Myron. **The Runaways: Children, Husbands, Wives and Parents.** Boston: Little, Brown, 1978. 239p.

The author looks at the phenomenon of running away, regardless of age. He tries to answer the questions about motives for running away, calling it a desperate act similar to suicide.

Connors, Patricia, with Dorianne Perucci. **Runaways: Coping at Home and on the Street.** New York: Rosen Publishing, 1989. 150p.

Written by the executive director of Covenant House, the primary institution dealing with teens on the street, this book examines some of the reasons why teens run away from home and provides coping techniques both for staying at home and for dealing with the terrifying possibilities on the street.

Janus, Mark-David, Arlene McCormack, Ann Wolbert Burgess, and Carol Hartman. **Adolescent Runaways: Causes and Consequences.** Lexington, MA: D. C. Heath, 1987. 154p.

A new study based on 149 runaway youths, ages 16 to 21, who sought refuge in Toronto, Canada, in the summer of 1984. The authors analyze the causes of running away, how runaways can be helped, and the general prognosis. One chapter consists of the interpretation of drawings by runaway youths.

Kosof, Anna. **Runaways.** New York: Franklin Watts, 1977. 111p.

Describes the life of runaway youths, situations they are trying to escape, and programs that try to help them solve their problems.

Paine, Roger W., III. **We Never Had Any Trouble Before.** New York: Stein & Day, 1975. 181p.

This is an excellent book, reviewing all aspects of the runaway problem. The author offers reasons for running away, ranging from parental abuse to feelings of powerlessness at home. Interestingly, he develops the idea that in today's homes with smaller families and both parents working, there are fewer safety valves to relieve pressures that children experience.

Raphael, Maryanne, and Jenifer Wolf. **Runaways: America's Lost Youth.** New York: Drake, 1974. 176p.

The stories of nine different teens who have run away are described in some detail. The observations of the authors are included in a brief introduction and conclusion frame.

Roberts, Albert R. **Runaways and Non-Runaways in an American Suburb.** New York: The John Jay Press, 1981.

This book offers a comprehensive look at the problem of runaways. The author reviews all of the research on the topic, concluding that running away means vastly different things to different children and changes with time and circumstance. There is no sharp image of the runaway available from the research.

Rubin, Arnold P. **The Youngest Outlaws: Runaways in America.** New York: Messner, 1976. 191p.

A discussion of runaways in the United States, including case studies.

U.S. Department of Justice. **America's Missing and Exploited Children: Their Safety and Their Future.** Washington, DC: U.S. Government Printing Office, 1986. 32p.

This report of the U.S. Attorney General's Advisory Board on Missing Children questions the attitude that parents and legal guardians should not interfere with a child's decision to leave home in search of an individual identity. It recommends actions that will help to prevent children from running away or being victimized.

U.S. Department of Transportation, Federal Railroad Administration. **On the Run: A Guide for Helping Runaway Youth in Transportation Centers.** Washington, DC: U.S. Government Printing Office, 1980. 28p.

This guide was published in response to the increasing number of runaways in the proximity of transportation centers. It outlines assistance programs and provides a brief directory of organizations and programs directly concerned with runaways.

U.S. Senate. **Exploitation of Runaways.** Washington, DC: U.S. Government Printing Office, 1986. 123p.

A report of a hearing before the Subcommittee on Children, Family, Drugs, and Alcoholism of the Committee on Labor and Human Resources that examines alternative ways to serve runaways and homeless youths.

U.S. Senate. **Exploited and Missing Children.** Washington, DC: U.S. Government Printing Office, 1982. 92p.

This report of a hearing before the Subcommittee on Juvenile Justice of the Committee on the Judiciary reviews the general concerns of the problems of juvenile delinquency and the impact on juveniles of the kinds of activity affecting later criminal conduct. Child pornography is the greatest concern here.

U.S. Senate. **Missing Children's Assistance Act.** Washington, DC: U.S. Government Printing Office, 1984. 278p.

This report of the hearing before the Subcommittee on Juvenile Justice of the Committee of the Judiciary deals with a bill to amend the 1974 Juvenile Justice and Delinquency Prevention Act by providing assistance in locating missing children. It includes publicity surrounding missing children and public response to such cases as those of Adam Walsh and Etan Patz.

U.S. Senate. **Private Sector Initiatives Regarding Missing Children.** Washington, DC: U.S. Government Printing Office, 1986. 74p.

Report of a hearing before the Subcommittee on Juvenile Justice of the Committee on the Judiciary that focuses attention on missing children—runaways, parental abductions, and kidnapping—and the responses of the private sector to this problem.

U.S. Senate. **Problems of Runaway Youth.** Washington, DC: U.S. Government Printing Office, 1983. 106p.

A report of a hearing before the Subcommittee on Juvenile Justice of the Committee on the Judiciary that discusses the adequacy of funding the runaway youth program through youth shelters.

Watson, Ian. **Double Depression: Schooling, Unemployment, and Family Life in the Eighties.** Sydney, Australia: Allen & Unwin, 1985. 161p.

Through case studies of runaway youth in Australia, the book addresses issues contributing to their problems.

Weisberg, D. Kelly. **Children of the Night: A Study of Adolescent Prostitution.** Lexington, MA: D. C. Heath, 1985. 298p.

A carefully documented study of the backgrounds and lifestyles of both male and female adolescent prostitutes. The connection between runaways and prostitution is explored.

ARTICLES

Berman, Claire. **"The Runaway Crisis,"** *McCall's* 115, no. 4 (January 1988): 113–116.

A high school girl who disappears with her older boyfriend is used as an example of runaway youths looking for independence and romance.

Hersch, Patricia. **"Coming of Age on City Streets,"** *Psychology Today* 22, no. 1 (January 1988): 28–37.

The outreach program at Covenant House is the setting for the runaway children in this article, presenting a grim picture of crack, prostitution, violent pimps, and AIDS.

Lonnborg, Barbara, and Robert B. McCall. **"Unlikely Runaways,"** *Parents* 59, no. 12 (December 1984): 180–184.

The articles attempts to help parents cope in the event of a runaway, but emphasizes that 20 percent of runaways are not alienated or delinquent youths.

"Runaway Youth: A Profile," *Children Today* 15, no. 1 (January/February 1986): 4–5.

A report on the youths served in 210 agencies, summarizing their characteristics and concluding that many are throwaways or pushouts from their homes.

Nonprint Materials

Daddy's Girl—Changing Family Relationships
Type: 3/4″ video
Length: 15 min.
Rent from: Indiana University Audio-Visual Center
Franklin Hall, Room 0001
Bloomington, IN 47405
(800) 551-8620 (for cost and arrangements)
Date: 1980

Jean tries to run away when her father attempts to end her relationship with a young farm helper.

48 Hours on Runaway Street
Type: Various video; color
Length: 49 min.
Cost: Purchase $357 (3/4″), $300 (1/2″)
Distributor: Carousel Film & Video
260 Fifth Avenue, Room 705
New York, NY 10001
(212) 683-1660
Date: 1989

Originally a CBS documentary, this film depicts runaways in New York City, San Diego, and Fort Lauderdale, Florida, facing violence and danger as they attempt to survive on the streets.

The Runaway Problem

Type: 16mm film, various video
Length: 13 min.
Cost: Rental $75 (three days)(plus shipping); purchase $315 (film), $220 (video)
Distributor: Coronet/MTI Film & Video
105 Wilmot Road
Deerfield, IL 60015
(800) 621-2131; (708) 940-3600
(call collect from Illinois and Alaska)
Date: 1980

Stories in which teenagers who have actually run away tell of their anguish and helplessness.

Runaways

Type: 16mm film, video
Length: 11 min.
Cost: Rental $20 per showing (film), $15 per showing (video); purchase $140 (film), $97.50 (video)
Distributor: Teleketics, Franciscan Communication Center
1129 South Santee Street
Los Angeles, CA 90015
(213) 746-2916 or (800) 421-8510
Date: 1985

A young boy runs away from home because he thinks he's "nobody important"—his parents don't seem to love each other or acknowledge him.

Streetwise

Type: Video
Length: 92 min.
Cost: Purchase $89.95
Distributor: Angelika Films
1974 Broadway
New York, NY 10023
(212) 769-1400 (for rental costs and arrangements)
Date: 1985

This film documents the lives of homeless young people in Seattle.

Street Shadows

Type: Various video; color
Length: 29 min.
Cost: Rental $75 (one day)(plus $5.75 shipping); purchase $149 (VHS or Beta), $199 U-Matic
Distributor: Films for the Humanities and Sciences
P.O. Box 2053
Princeton, NJ 08543
(800) 257-5126
Date: 1985

Here the stories of throwaways and runaways are told by the kids themselves. It's not a pretty picture they describe as they reveal how they get their money, how they survive, and, sometimes, how they die.

Organizations

Children of the Night
1800 N. Highland Avenue, 128
Hollywood, CA 90028
(213) 461-3160
Executive Director: Dr. Lois Lee

Aims to provide protection and support for young people ages 8–17 who are involved in street life by putting them in contact with referral services, outreach programs, and walk-in centers. Currently concentrating on the Los Angeles area, but accepts requests for information from all parts of the United States.

Homelessness Information Exchange
1830 Connecticut Avenue NW
Washington, DC 20009
(202) 462-7551
Executive Director: Dana Harris

Provides information about homelessness.

PUBLICATION: *Homewords*, quarterly.

National Network of Runaway and Youth Services
905 Sixth Street SW, Suite 411
Washington, DC 20024
(202) 488-0739
Executive Director: June Bucy

Acts as an information clearinghouse for community-based human services agencies dealing with the concerns of runaways and other troubled youth.

PUBLICATIONS: *Network News*, monthly; *Policy Reporter*, monthly.

Hotlines

Boys Town
(800) 448-3000

This hotline refers males and females to shelters in most metropolitan centers and makes transportation arrangements for stranded runaways who wish to return home.

Covenant House
(800) 999-9999

Callers to the Covenant House "nine line" can receive referrals to shelters and, when possible, counseling resources in the area from which they have called.

National Runaway Switchboard
2210 North Halsted
Chicago, IL 60614
(800) 621-4000
Toll-free number in Illinois: (800) 972-6000.
Executive Director: Baulkus Heard

A 24-hour, toll-free national switchboard for runaways, families of runaways, and other troubled youth. Provides names, addresses, and phone numbers of centers for shelter and other social services across the country, including counseling centers, referral lines, drug treatment facilities, and family planning services. Offers to relay messages between young people and their families if desired; can also set up conferences between youth and parents or agencies. The caller's confidentiality is maintained. Funded in part by the office of Youth Development of the U.S. Department of Health and Human Services.

CHAPTER 5

Teen Crimes: Violence against Persons

Roger slammed the ball into the side of his face. There was a sickening thud of rubber against flesh and a muffled cry from the boy before he toppled over backward.

"Big basketball player," Roger said. "Can't even catch."

The boy rolled onto his side slowly and tried to push himself up, each movement deliberate and painful. His arms strained and his body began to rise. He was about halfway up when he faltered, his arms gave way under the weight, and he collapsed face down on the ground.

. . . For several seconds the guys tried to decide if Roger had merely knocked him out, or was the guy seriously injured—dead, or what?

Sticks rolled the boy so he was face up. Instead of the slight movement of before, his head jerked from side to side, eyes shut but mouth wide open, sucking in air noisily. The twitching stopped suddenly, but immediately his body contorted grotesquely, one leg bending and flopping to the side while his left hand reached straight out as if to grab something or someone. The boy's back arched and stiffened and a wet, gurgling sound escaped from deep in his throat.

Jim Murphy, *Death Run* (New York: Clarion, Houghton Mifflin, 1982), 11, 12.

The teenager in Jim Murphy's *Death Run* was dead. The guys had decided to play a prank on the teenager in the park. They didn't know him, but he had come to the park to shoot a few baskets, and that made him a despised jock. Taking the basketball away from him was easy. After a little keep-away, the jock became angry and called them bastards. What started as a simple prank became violence, then murder, or maybe manslaughter if they got a good lawyer to defend them in court.

In Larry Bograd's novel, *Bad Apple*, readers are witness to the brutal beating of an old couple rumored to have a lot of cash hidden in a safe in their mansion. Prune and Nicky decide to rip them off. But after breaking into the house and confronting the terrified couple, it is plain to see that there is no safe. And no money. Fifteen-year-old Nicky flies into a rage.

> So I pick—rather, I watch myself pick up—a heavy stone ashtray. Somewhere from my gut comes a sound I didn't know I could make, a yelp, as I glide to the couple, frozen there like two dumb animals, and start to hit them. My arm, swinging back and forth, feels free and easy. All the tension is gone.
>
> Larry Bograd, *Bad Apple*
> (New York: Farrar, Straus & Giroux, 1982), 144.

Statistics on Teen Violence against Persons

- In a survey of university men, 10 percent admitted forced sexual intercourse and another 50 percent admitted to some kind of coercive sexual behavior (McDonald, 59).
- Of all those arrested for murder in 1987, 44 percent were under 25 years of age, and 10 percent were 17 or younger (FBI, 12).
- In one survey, about half of all rapes reported around the country involved teenagers, either as victims or offenders (Curtis, 1974).
- Aggravated assault arrests, including those of teens, are up sharply over previous years—in 1987 the rate was 26

percent higher than in 1983 and 34 percent higher than in 1978 (FBI, 22).

Definitions

"**Rape** is forced copulation of one person by another. By forced copulation we mean copulation without explicit or implicit consent; it need not involve physical force" (Thornhill, Thornhill, and Dizinno, 113).

"**Murder and nonnegligent manslaughter** . . . is the willful (nonnegligent) killing of one human being by another" (FBI, 7).

"**Aggravated assault** is an unlawful attack by one person upon another for the purpose of inflicting severe or aggravated bodily injury" (ibid., 21).

The Scope of Teen Violence against Persons

Violence against others may have become an integral part of American culture. At least it has become part of our entertainment, some of our music, daily TV news fare, and, more frequently, the means for resolving family conflicts. It is a fact that crimes of violence have been on the increase nationally over the past decade (FBI, 8–23). Rapes, assaults, and homicides involving teenagers have also been increasing over the years.

In the early 1980s, Milpitas, California, townspeople were shocked to learn the details of the callous murder of 14-year-old Marcy Conrad by her 16-year-old boyfriend. He had abandoned the body in a rural area outside of town, then talked openly with his school friends about the murder. Unbelievably, none reported the murder to the police. "Over the next two days nearly a dozen teenagers traipsed into the hills to peer at Marcy's half-nude body. One gawker dropped a rock on her face to see if the corpse was real; others poked at it with sticks" (Abramson, 25).

Most who learned of the details were not only repulsed by the violence itself, but also were at a total loss to explain why the teenage friends of the victim and her murderer felt no remorse, no responsibility, and no need to get involved by informing the police. In *River's Edge*, a movie based upon the case, two of the high school friends of the couple are shown in a moment of sober reflection upon the crime, wondering about their own feelings. Following is a brief excerpt of their dialogue:

> Matt: "It affected me—didn't it affect you? You knew Jamie [the victim], right? And there she was—dead—right in front of us. And even that close, we don't even feel that we've lost anything."
>
> Clarissa: "I didn't even cry for her."

Homicide

Violent homicides are not unique to California, suburbs or otherwise. In Detroit's Murray Wright High School, fellow students were shocked to see Charles Jackson, Jr., a star football player, gunned down in cold blood. His 14-year-old murderer had chased him down the hallways firing a .357 magnum pistol, wounding two other students, and terrorizing the entire school (Wilkerson, A-1).

In the Dallas suburb of Midlothian, Texas, a young undercover narc in the local high school was shot to death along a lonely country road, and two teenagers accused of the murder will be tried as adults. One of the two—16-year-old Greg Knighten—is the son of a veteran Dallas policeman who, ironically, was the one who took the call of a tipster that led to the arrest of the two teens. Neither teen had any serious record of prior offenses, and the school and the community had just begun to talk about how to cope effectively with what they believed was a relatively minor drug problem (Kennedy, 7).

In Washington, D.C., 15-year-old Sean Smith was gunned down while involved in an argument over his red ski jacket, which he had just bought with earnings from his part-time job (Gaines-Carter, A-1).

In Gallup, New Mexico, two teenagers were accused of the murder of a test flight engineer who was scheduled to consult with local Indians about an aviation training program he proposed to start. His body was found stuffed in the trunk of his car at a lonely location six miles outside of town. Police were reported to believe that robbery was the motive (Minicler, 1-A).

A 16-year-old in Chicago shot and killed a motorist who had stopped at an alley, and when brought to trial was released because the witnesses failed to show up (Warner, 18). The stories of violence are almost endless, and the teens who are involved in these senseless crimes (except for the gang wars) appear to have little in common.

Rape

Rape is another common form of violence against persons. Although it is a sexual crime, it is considered today to be a crime of violence more than an act of unrestrained sex (Tobach and Sunday, 1985). Guys sometimes talk about a girl who "was asking for it," or the common notion that she did something to lead him on. But the fact remains that it is rape unless the partner freely and willingly gives her consent. It is true that males are sometimes the victims of rape, particularly teenage boys who may be imprisoned with homosexual adult males. But, the vast majority of victims are female, and virtually always the offender is male.

Teens are likely to be involved in rape, either as victim or rapist. One researcher reported that the percentage of teenage girls who say that they were raped ranges from 7 to 9 percent. However, this means that overall in the United States the number of teens raped might range from 600,000 to as high as 1.5 million in one year (Ageton, 26).

Although the concept of "date rape" may be fairly new, there is now increasing attention to rape committed by males who are friends, or at least acquaintances, of the victims. "Most adolescent male offenders seemed to be responding to a situation, usually a date, that afforded an opportunity to pursue sexual interests" (ibid., 95). Date rapes occur almost anywhere, but the research indicates the high likelihood of rape in the victim's home, the offender's home, or in an automobile.

Offenders report that the type of force used is usually verbal persuasion, though drugs and alcohol frequently play a part (ibid., 96).

Unfortunately, males often view the act of rape as more of a seduction or sexual encounter with a hesitant partner than as an assault. One researcher who studied college sophomore males reported that 75 percent of them admitted to having used drugs or alcohol to have sex with their date, and 69 percent acknowledged using verbal persuasion. Further, this researcher concluded that these men's "macho values" produced in them "greater tolerance for rape and sexual exploitation" (Mosher, 12). Again, to repeat a point, rape is aggression, and when anyone has sex with an unwilling partner, they are committing a crime.

The sheer brutality of the crime of rape is illustrated by the highly publicized attack by a group of teenagers upon a female jogger in New York's Central Park in the spring of 1989. Thirty-some teenage boys had entered the park about 9 o'clock in the evening to engage in random acts of assault that they called "wilding" (Gelman, 65). The gang split into smaller groups as their rampage proceeded. Apparently their initial targets were cyclists and joggers, though they robbed a 52-year-old man along the way. Wielding pipes and throwing rocks, they climaxed their crime spree in an attack on a 29-year-old female investment broker who was out for her usual evening run (Gibbs, 20–21). One account said that "she screamed and struggled as they dragged her 200 feet down the ravine, throwing her to the ground at the edge of a pond. They tore her clothes off and bound her hands and gagged her with her sweatshirt. While some held her, others—perhaps four of them—raped her" (Kunen, 108). During the rape, her face and body were beaten with stones and a piece of metal pipe, and she was left, bleeding and unconscious (ibid.).

Several hours later she was discovered by several hikers, but in the meantime she had lost 80 percent of her blood supply, and had slipped into a coma. Six teenagers were indicted for the rape and stood trial as adults, though their ages ranged from 14 to 16. Fourteen months later, when the trial began, the victim still had no memory of the attack (Sullivan, B-1).

Whether a vicious brutal offense such as took place in Central Park, or the increasingly common date rape, rape is a major crime of violence.

REFERENCES

Abramson, Pamela. "Bitter Memories of Murder," *Newsweek* 109, no. 25 (11 June 1987): 25.

Ageton, Suzanne. *Sexual Assault among Adolescents*. Lexington, MA: Lexington Books, 1983.

Bograd, Larry. *Bad Apple*. New York: Farrar, Straus & Giroux, 1982.

Curtis, L. A. *Criminal Violence*. Lexington, MA: Lexington Books, 1974.

Federal Bureau of Investigation. *Uniform Crime Reports of the United States*. Washington, DC: U.S. Department of Justice, 1988.

Gaines-Carter, Patrice. "Don't Let Sean's Death Be In Vain," *Washington Post*, 5 January 1988: A-1.

Gelman, David. "Going 'Wilding' in the City," *Newsweek* 113, no. 19 (8 May 1989): 65.

Gibbs, Nancy. "Wilding in the Night," *Time* 133, no. 19 (8 May 1989: 20–21.

Kennedy, J. Michael. "Death of a Young Narc: Town Searches for Motive," *Los Angeles Times*, 23 November 1987: 7.

Kunen, James S. "Madness in the Heart of the City," *People's Weekly* 31, no. 20 (22 May 1989): 108.

McDonald, Kim. "Sex under Glass," *Psychology Today* 22, no. 3 (March 1988): 53.

Minicler, Kit. "Denver Man Found Dead; Two Teens Held," *Denver Post*, 29 July 1989: 1-A.

Mosher, Donald L. "Rape: The Macho View," *Psychology Today* 21, no. 4 (April 1987): 12.

Murphy, Jim. *Death Run*. New York: Clarion, Houghton Mifflin, 1982.

Sullivan, Ronald. "Participants May Testify in Jogger Case," *New York Times*, 14 June 1990: B-1.

Thornhill, Randy, Nancy W. Thornhill, and Gerard Dizinno, "The Biology of Rape," in *Rape*, ed. Sylvana Tomaselli and Roy Porter (New York: Basil Blackwell Ltd., 1985).

Tobach, Ethel, and Suzanne Sunday, eds. *Violence against Women*. Staten Island, NY: Gordian Press, 1985.

Warner, Edwin. "The Youth Crime Plague," *Time* 110, no. 2 (11 July 1977): 18–28.

Wilkerson, Isabel. "Detroit Crime Feeds on Itself and Youth," *New York Times*, April 29, 1987: A-1.

Resources
for Finding Out about Violence against Persons

Fiction

Ashley, Bernard. **A Kind of Wild Justice.** New York: S. G. Phillips, 1979. 182p.

Ronnie knew that he and his father were at the mercy of the vicious Bradshaw brothers. The Bradshaws were well known to the local police and had kept the neighborhood in terror for several years. As this novel roars to a conclusion, justice wins out!

Bograd, Larry. **Bad Apple.** New York: Farrar, Straus & Giroux, 1982. 152p.

Unwanted, abused, and blamed for the death of his small sister, Nicky joins in crime with Prune. Felony leads to assault as the two boys rob and beat an elderly couple, leading to Nicky's arrest.

Cormier, Robert. **Beyond the Chocolate War.** New York: Knopf, 1985. 278p.

Problems between students escalate at the exclusive, private Trinity High School. Archie is still running the school, in spite of the vicious beating of dissenter Jerry Renault, but Archie's right-hand man, Obie, is starting to seek his own power.

Duncan, Lois. **Killing Mr. Griffin.** New York: Dell/Laurel, 1979. 224p.

Four teens, led by unstable Mark and aided by shy Susan, kidnap their despised English teacher, whose bad heart causes his death. The students try to cover up what happened, but Susan's guilty conscience gives them away.

Guy, Rosa. **The Disappearance.** New York: Delacorte, 1979, 246p.

Imamu Jones has had a manslaughter charge against him dismissed. Because of his poor home situation with an alcoholic mother, he is put into a foster family. When the youngest child in the family, Perk, is missing, Imamu is suspected.

Levy, Marilyn. **Putting Heather Together Again.** New York: Ballantine/Fawcett Juniper, 1990. 186p.

When Heather's breakup with her boyfriend leaves her emotionally vulnerable, Joe, an older guy, takes advantage of her and rapes her. Heather visits a rape crisis center and learns what it takes to recover.

Miklowitz, Gloria. **Did You Hear What Happened to Andrea?** New York: Delacorte, 1979. 168p.

Fifteen-year-old Andrea is hitchhiking home from the beach when she is raped. Her assailant is caught, but she must deal with the reactions of her boyfriend, friends, and family.

Miller, Frances A. **The Truth Trap.** New York: Dutton, 1980. 248p.

After his parents' death, 15-year-old Matt flees to Los Angeles, and is then accused of the brutal murder of his younger sister. He is finally cleared, but has to overcome his reputation.

Murphy, Jim. **Death Run.** New York: Clarion, Houghton Mifflin, 1982. 174p.

Brian and his friends steal a six-pack of beer and assault a boy in the park. The boy dies, although not because of the assault. Brian's guilt causes him to run, even though the daughter of the detective investigating the crime helps him. Finally his innocence is established.

Peck, Richard. **Are You in the House Alone?** New York: Dell/Laurel, 1979. 176p.

Gail is being harassed by notes and calls. When she is finally confronted and raped by a fellow student, she has to face the trauma of reporting her classmate, who is from an influential family.

Shaw, Diana. **Lessons in Fear.** New York: Little, Brown, 1987. 172p.

An unpopular ninth-grade teacher is found unconscious, and the evidence of the assault points towards popular student Adrian. Carter manages to overcome her personal problems with her divorced mother to find the real assailant.

Silsbee, Peter. **The Big Way Out.** New York: Bradbury, 1984. 180p.

Peter and his mother flee from the tension of an abusive home life in which his mentally ill father has convinced his brother Tim that everything is normal. In order to end the nightmare, Peter decides to shoot his father.

Nonfiction

Adams, Caren, Jennifer Fay, and Jan Loreen-Martin. **No Is Not Enough: Helping Teenagers Avoid Sexual Assault.** San Luis Obispo, CA: Impact Publishers, 1984. 192p.

Many practical suggestions for teens are presented in this paperback. Every chapter is relevant, but Chapter 10, "Love and Work: Avoiding Abuse in Relationships," is really a must for teenagers, both male and female.

Beneke, Timothy. **Men on Rape.** New York: St. Martin's, 1983. 192p.

Men's basic attitudes about rape are formed early in life and often crystalize during the teen years. In this volume, Beneke interviews all kinds of men of varying ages—some are rapists, some are not—who express a wide range of attitudes towards the act of rape.

Carpenter, Cheryl, Barry Glassner, Bruce D. Johnson, and Julia Loughlin. **Kids, Drugs, and Crime.** Lexington, MA: Lexington Books, D. C. Heath, 1988. 256p.

A comprehensive account of the relationship of crime to drug use among teenagers. Chapter 6, "The Context of Violence," provides relevant background material on violence among teens.

Estrella, Manuel, and Martin Frost. **The Family Guide to Crime Prevention.** New York: Beaufort Books, 1981. 253p.

As the title suggests, this book addresses family members in general about prevention of violent crime against themselves. The volume covers home security as well as practical measures to take to avoid becoming a victim of street violence. The chapter on avoiding rape and sexual assault is very comprehensive.

Fein, Judith. **Are You a Target?** Duncan Mills, CA: Torrance Publishing, 1988. 146p.

Based upon an earlier book by Fein (published by Wadsworth), this paperback offers female readers hard-hitting practical advice on how to avoid becoming a victim. Although experts aren't in total agreement on the overall effectiveness of fighting back, Chapter 3 on self-defense methods is very thorough and quite convincing.

Grossman, Rochel, and Joan Sutherland, eds. **Surviving Sexual Assault.** New York: Congdon & Weed, 1983. 96p.

A practical manual on what to do following a sexual assault. Lots of tips here, ranging from dealing with guilt to paying for the costs of medical treatment. The book even has a small section for male victims of homosexual assault.

Neiderbach, Shelley. **Invisible Wounds: Crime Victims Speak Out.** New York: Harrington Park Press, 1986. 200p.

This account of the plight of victims of violent crime presents interview material from victims with accompanying passages giving legal and psychological explanations. The juxtaposition of these two types of information is effective, and the author covers the emotional scarring of victims thoroughly.

Reiff, Robert. **The Invisible Victim.** New York: Basic Books, 1979. 219p.

Psychologist Reiff makes the point in this book that our justice system typically doesn't do much to help the victims of violent crime. The last three chapters strike a positive note by discussing what we can do about it.

Sanders, William B. **Juvenile Delinquency.** New York: Holt, Rinehart & Winston, 1981. 420p.

This is a well-known, basic text on the broad subject of delinquency. Chapter 9, "Delinquent Gangs," contains an excellent section on violence and weapons.

Tobach, Ethel, and Suzanne Sunday, eds. **Violence against Women.** Staten Island, NY: Gordian Press, 1985. 364p.

An annotated bibliography that offers resources for anyone doing research on rape and sexual abuse, including sexual harassment.

Wolbert, Ann, ed. **Rape and Sexual Assault.** New York: Garland Publishers, 1988. 215p.

A basic reference with an especially good section dealing with the social context of rape. Wolbert includes a chapter on the sexual abuse of boys—an often overlooked subpopulation of victims.

Nonprint Materials

Acquaintance Violence

Type: Video, color
Length: 15 min.
Cost: Rental $75; purchase $320
Distributor: Coronet/MTI Film & Video
108 Wilmot Road
Deerfield, IL 60015
(800) 621-2131; (708) 940-3600
(call collect from Illinois and Alaska)
Date: 1989

The young man pictured in this documentary was a former football star who was sentenced to prison for the murder of a co-worker. The film offers an explanation of violence as a learned behavior in response to conflict situations, and shows how alternative responses are more productive.

Aftershock: The Victims of Crime

Type:	16mm film, various video; color
Length:	25 min.
Cost:	Rental $75; purchase $520 (film), $470 (video)
Distributor:	Coronet/MTI Film & Video 108 Wilmot Road Deerfield, IL 60015 (800) 621-2131; (708) 940-3600 (call collect from Illinois and Alaska)
Date:	1982

Victims of rape and assault recall their trauma in this excellent documentary. Both victims and their families suffer indescribable anxieties from these violent crimes.

Against Her Will

Type:	Video, color
Length:	46 min.
Cost:	Rental $75; purchase $495
Distributor:	Coronet/MTI Film & Video 108 Wilmot Road Deerfield, IL 60015 (800) 621-2131; (708) 940-3600 (call collect from Illinois and Alaska)
Date:	1990

Acquaintance rape on college campuses is an increasing problem. This documentary is hosted by Kelly McGillis and examines the dynamics of this form of violence that faces many teens.

Andrea's Story: Hitchhiking Tragedy

Type:	16mm film, various video; color
Length:	30 min.
Cost:	Rental $90; purchase $630 (film), $475 (video)

Distributor: Coronet/MTI Film & Video
108 Wilmot Road
Deerfield, IL 60015
(800) 621-2131; (708) 940-3600
(call collect from Alaska and Illinois)

Date: 1984

Hitchhiking is still popular with many teens, and this award-winning film shows Andrea faced with a rapist while hitching a ride home. Her feelings following the ordeal are carefully and sensitively portrayed, and teens will feel their impact.

Beyond Rape

Type: 16mm film, various video; color

Length: 28 min.

Cost: Rental $70; purchase $520 (film), $415 (video)

Distributor: Coronet/MTI Film & Video
108 Wilmot Road
Deerfield, IL 60015
(800) 621-2131; (708) 940-3600
(call collect from Illinois and Alaska)

Date: 1984

A general treatment of several aspects of sexual violence, this film offers good ideas on rape prevention. There is some time spent on examining the motives of sex offenders.

Combat in the Classroom

Type: Video; color

Length: 27 min.

Cost: Purchase $420

Distributor: Coronet/MTI Film & Video
108 Wilmot Road
Deerfield, IL 60015
(800) 621-2131; (708) 940-3600
(call collect from Illinois and Alaska)

Date: 1980

Teachers are frequently assaulted by students, and this film, narrated by Ed Asner, discusses the spread of this violence. Teachers, students, and administrators are interviewed.

No Means No! Avoiding Date Abuse

Type: 16mm film, various video; color
Length: 18 min.
Cost: Rental $75; purchase $465 (film), $365 (video)
Distributor: Coronet/MTI Film & Video
108 Wilmot Road
Deerfield, IL 60015
(800) 621-2131; (708) 940-3600
(call collect from Illinois and Alaska)
Date: 1988

Boys viewing this film are reminded that they don't have to abuse females to prove they are men. The film depicts Lisa and her teenage boyfriend in a familiar situation—he wants sex, and she says no. Girls will learn from this film that their bodies are their own and that they can say no to unwanted sex anytime, anywhere.

Richie

Type: 16mm film, various video; color
Length: 31 min.
Cost: Rental $75; purchase $525 (film), $370 (video)
Distributor: Coronet/MTI Film & Video
108 Wilmot Road
Deerfield, IL 60015
(800) 621-2131; (708) 940-3600
(call collect from Illinois and Alaska)
Date: 1978

Robby Benson and Ben Gazarra are the stars of this excellent film about a rebellious 16-year-old boy who gets into drugs. He experiences family conflict and dies a violent death at the hands of his father. This film links violence to dysfunctional families.

Someone You Know: Acquaintance Rape

Type: 16mm film, various video; color
Length: 29 min.
Cost: Rental $85; purchase $520 (film), $470 (video)

Distributor: Coronet/MTI Film & Video
108 Wilmot Road
Deerfield, IL 60015
(800) 621-2131; (708) 940-3600
(call collect from Illinois and Alaska)
Date: 1986

Another award-winning film, featuring interviews with victims of acquaintance rape as well as with the convicted offenders. The contrast is apparent. Males maintain that the victim wanted, or owed them, sex. For the female victims, the feelings are guilt and pain.

Sugar and Spice and All Is Not Nice
Type: 16mm film, various video
Length: 19 min.
Cost: Rental $75; purchase $415 (film), $345 (video)
Distributor: Coronet/MTI Film & Video
108 Wilmot Road
Deerfield, IL 60015
(800) 621-2131; (708) 940-3600
(call collect from Illinois and Alaska)
Date: 1984

This film examines the social and cultural context in which rape occurs. Opinions about what causes crimes against women are offered by rape victims and counselors. Viewers gain a greater understanding of the factors in our society that influence brutal rapes.

Teen Violence
Type: Video, color
Length: 28 min.
Cost: Rental $75; purchase $395
Distributor: Coronet/MTI Film & Video
108 Wilmot Road
Deerfield, IL 60015
(800) 621-2131; (708) 940-3600
(call collect from Illinois and Alaska)
Date: 1989

Teenagers in the inner city face violence daily, and this film examines the causes as well as the means that most teens use to avoid the violence. The prevailing theme in this film is that teen violence is a function of poverty rather than race, and that those who live in these inner-city areas have often learned how to prevent violence as well as live with it.

Violence: Will It Ever End?
Type: 16mm film; color
Length: 19 min.
Cost: Not given
Distributor: Document Associates/The Cinema Guild
1697 Broadway
New York, NY 10019
(212) 246-5522
Date: 1978

This film examines root causes of violence in our society. Rollo May, noted psychiatrist, discusses the problem, along with the underlying personality dynamics of offenders.

Organizations

Children's Defense Fund
122 C Street NW
Washington, DC 20001
(202) 628-8787
President: Marian Wright Edelman

Provides advocacy for children in many areas, including criminal defense.

PUBLICATION: *CDF Reports*, monthly.

Conflict Resolution/Alternatives to Violence
P.O. Box 256
Ricker House
Cherryfield, MA 04622
(207) 546-2780
Correspondence Secretary: William Conway

Offers a referral service for resources on training in conflict resolution.

National Coalition against Sexual Assault
2428 Ontario Road NW
Washington, DC 20009
(202) 483-7165 (national office)
President: Mary Beth Carter
(415) 236-7273

A network providing referral to state and local resources and hotlines for victims of rape and sexual assault.

PUBLICATION: *Newsletter*, quarterly.

National Organization for Victim Assistance
1757 Park Road NW
Washington, DC 20010
(202) 393-6682
Executive Director: Marlene A. Young

Offers counseling and advocacy services to victims, as well as referral to a network of services for teens and adults in 50 states.

PUBLICATION: *NOVA Newsletter*, quarterly.

WE CARE
Route 2, Box 33M
Altmore, AL 36502
(205) 368-8818
Director: Phil Weber

Interested in deterring teens from a life of crime. Sends volunteers into detention facilities to talk about changing from criminal behavior.

PUBLICATION: *We Care Newsletter*, every 4–6 weeks.

Hotlines

Check local social service agencies for crisis lines serving your region.

CHAPTER 6

Teen Crimes: Violence against Property

> The sales clerk walked behind the counter towards Arnold. He leaned down to open the case holding the charm bracelets. Moving his hand so quickly I could scarcely see it, Stewie snatched one of the watches from the tray—not one of the two between which he was supposedly making his choice, but another one, a more expensive one, made in Switzerland. He dropped the watch into his pocket.
>
> No sooner had the watch disappeared from view than a loud bell rang throughout the store. I was so startled by the sound that my whole body shivered at once. Stewie drew his breath in sharply. From the workroom in back of the store, Mr. Lippman himself hurried into view. The clerk stood up, a satisfied smirk on his face. "I caught them, Mr. Lippman," he shouted. "I caught them red-handed."
>
> Barbara Cohen, *King of the Seventh Grade* (New York: Lothrop, Lee & Shepard Books, 1982), 119.

Thirteen-year-old Vic Abrams and his buddies had stolen before, but this was the first time they had been caught at the scene of the theft.

> Christmas Eve, Officer Muncie caught him carrying a case of wine out of a liquor store in broad daylight. How could Jason have believed he wouldn't get

> caught? A minor, strutting down Main Street with twelve bottles of rose? I guess he knew he wouldn't get away with it; he didn't really want to.
>
> Jan Greenberg, *Bye, Bye, Miss American Pie* (New York: Farrar, Straus & Giroux, 1985), 147.

In Jan Greenberg's book, Jason is pictured as good-looking and likeable, yet cocky and wild. He just wants to show off, and he is caught.

In Jim Murphy's novel, *Death Run,* Brian has just stolen a six-pack from a neighborhood convenience store and taken it out to the park to meet his buddies.

> He leaned over the fence, pulled the six-pack out hurriedly and broke a can from the plastic carrying band. "They're sixteen ounces, too," he said, handing one to Roger. "MacPheason leaves them near the door."
>
> "You mean that old coot didn't scream murder?" Roger tore the aluminum tab open and a jet of foam shot into the air, swirled, and landed on the front of Stick's jacket. Roger flashed Brian a quick smile and took a long gulp before handing the can to Al.
>
> "I waited until he went into the back room," Brian explained. "He watches game shows most of the time, so it was easy."
>
> Jim Murphy, *Death Run* (New York: Clarion, Houghton Mifflin, 1982), 4.

Statistics on Teen Property Crimes

- Of those arrested recently for larceny/theft, 33 percent were under the age of 18 years (FBI 1987, 8)
- In 1987, 40 percent of those arrested for arson were 18 years or younger, and the vast majority were male (FBI 1987, 39)

- People under 18 have a higher likelihood of being arrested for robbery and other property crimes than any other age group (U.S. Department of Justice, *Report to the Nation,* 43)
- According to a recent survey, 86 percent of all shoplifters who were caught were female, and the vast majority of them were under 18 years of age (Sanders, 132)
- Among 1988 high school seniors, 37 percent of the boys and 24 percent of the girls admitted that they had taken something from a store without paying for it (U.S. Department of Justice, *Sourcebook,* 18)
- Among 1988 high school seniors, 20 percent of the boys and 8 percent of the girls admitted that they had damaged school property on purpose (ibid., 340–41)
- Among 1988 high school seniors, 17 percent of the boys and 13 percent of the girls admitted hurting someone badly enough to need medical attention (ibid.)

Definitions

Burglary. Entering a structure illegally to commit theft or other serious crime.

Larceny. In general, theft; "the unlawful taking, carrying, leading, or riding away of property from the possession or constructive possession of another" (FBI 1987, 28).

Robbery. Taking something of value from another person by force or threat of force.

Motor vehicle theft. Taking autos, trucks, buses, motorcycles, or snowmobiles that belong to another person; often referred to as joyriding (ibid., 30).

Arson. "Any willful or malicious burning or attempt to burn, with or without intent to defraud, a dwelling house, public building, motor vehicle or aircraft, or personal property of another" (FBI 1985, 36).

Vandalism. "Destroying or damaging, or attempting to destroy or damage, the property of another without his or her consent." (U.S. Department of Justice, *Report to the Nation*, 3).

Scope of Property Crimes

Each year thousands of teenagers are involved in property crimes. In addition to burglary, the most serious of these crimes, teens are arrested for shoplifting, minor thefts, car and bicycle thefts, arson, and vandalism. Nationwide, two out of three burglaries are in homes. Overall, those under age 18 account for nearly 20 percent of all burglaries and 40 percent of all motor vehicle thefts (FBI 1988, 27–41). Economic losses run into the millions of dollars, and no one has been able to assess the emotional costs to victims.

Perhaps the most tragic aspect of these teenage property crimes is the fact that for many they mark the beginning of a lifelong crime career. In one study of adult robbers, it was found that most of them had been contacted by police prior to age 18, and most often the contact was in connection with larceny (Roeback, 97–103). Evidence shows that teenage crimes are frequently the training ground for adult criminals (Sanders, 47–59). A former U.S. attorney general has observed that

> . . . youth is a time when character is forming, a time when an individual begins to know himself. It is difficult for many adolescents who see hunger and deprivation around them to understand why property should not be taken, or why sensitivity to the feelings of others is necessary. They do not yet understand.
>
> (Clark, 240)

Minor thefts, especially those involving teens, often go unreported; for example, such a common occurrence as a theft from a school locker. Suppose that someone took a raincoat from an open locker on a rainy day. If it were returned in another day or two, the incident would probably go unreported.

Even if the stolen object were not returned to the school locker, the likelihood of its being reported to school authorities, let alone to law enforcement agencies, would be very small. Thus, it is clear that thousands of petty thefts never become a part of the FBI's national statistics.

Among teenagers, shoplifting is the most common form of larceny. It is the type of theft that is increasing most rapidly from year to year—up 19 percent in a one-year period (FBI 1987, 30–31). It is also the type of theft that is the least reported. Estimates of losses to retailers have been placed at $26 billion a year, with a resulting increase in retail prices of from 5 to 7 percent (Reid, 267). Not only are large numbers of shoplifting incidents unreported, but also the public is notably apathetic. When a group of high school students staged a number of shoplifting incidents in connection with a marketing class (and in cooperation with the store manager), it was reported that many persons witnessed the crimes, but none reported them (Reid, 267).

Why Teens Commit Property Crimes

IMPULSE OR HABIT

Although much of the theft reported to police is linked to drugs, the reasons why teenagers first become involved in property crime are many and varied. Some do it for a lark. Others don't consider the eventual consequences of an impulsive act. A recent case reported in New York City involved three teens who had just graduated from Fordham Preparatory School. When a minor accident damaged a car that one of the boys had borrowed, they decided to go on a robbery spree to obtain money for the $900 worth of repairs required. They held up a grocery store, then attempted to rob a couple parked in a car. In the shooting that followed, an off-duty officer fired at the robbers and killed 18-year-old James Cooney, who was viewed by friends as a smart kid who talked about going to college (*New York Times*, A-1).

Not all who begin stealing end up in the morgue. In fact, the majority of teenagers who steal are prime candidates for lifelong careers of crime. Former Attorney General Ramsey Clark expressed it clearly:

> Young people who steal will find it as hard to change that habit as any other they acquire. When stealing is frequent it becomes a part of the style of living. Ostracized by society for conduct it condemns, the individual turns against society. It is human to defend one's own actions. As the protest of society is voiced, the conduct becomes more social. For some, society itself becomes the enemy.
>
> (Clark, 240)

ECONOMIC GAIN

Property crimes such as theft are usually for economic gain. Arson and vandalism are harder to understand, because they seldom result in any financial gain. An exception might be arson that is intended to give a property owner a windfall of insurance money. Larceny thefts are by far the largest number of offenses that are reported to law enforcement agencies. Items stolen are usually goods that can be converted easily into cash, e.g., jewelry, televisions, VCRs, stereos, furs, and cars. The cash may be needed to support a drug habit, or simply to provide for needs which go unmet in low-income or poverty homes.

Shoplifting represents both an attempt to satisfy an economic need and an act of impulse—a lark. More females than males are arrested for shoplifting (Flowers, 120). Why there are gender differences is not clear, and there are many possible reasons suggested by various writers ("Pilfering Urges," 94). In any case, the data seem to indicate that the differential rates for males and females involved in shoplifting may be narrowing (Sichor and Kelly, 75–76).

There are many individual motives for shoplifting. Experts generally separate those involved into professionals and amateurs. The amateurs are usually referred to as "snitches," and the professionals are called "boosters" (Cameron). Snitches usually take small items for personal use rather than for resale, and the items are often luxuries, or at least nonessential. They shoplift on impulse, or on a dare.

Shoplifters differ from other thieves in one important way—they usually do not think of themselves as thieves (Cameron). Arrest by authorities requires them to face up to the nature of the offense, but the research indicates that even after arrest they somehow don't view themselves as real thieves. Because store policies vary considerably from one retailer to another, it is likely that many shoplifters are allowed to avoid arrest records when restitution is offered or parents intervene with promises of parental discipline. Thus the failure to arrest and prosecute may have the unintended effect of allowing the shoplifter to deny personal responsibility and guilt. "The composure of juveniles being detained has never ceased to amaze me, that is until notified that they must tell a parent of their misdemeanor. Then the tears flow and pleadings begin" (Edwards, 135–136).

Often there is no need for the item that is stolen, so the motivation for the act lies elsewhere. "Shoplifters who have no physical need for the merchandise they steal may symbolically be trying to take what life has not given them: love, affection, and nurturing," says psychologist Joseph Smith ("More Than Sticky Fingers," 10). Amateur teenagers usually tuck the item into a pocket, purse, or backpack, in marked contrast to the professional boosters, who use false-bottomed boxes, baggy bloomers, and a variety of expandable garments designed to conceal the stolen item (Cameron, 39–49).

Theft for gain can begin at a very early (preteen) age. One recent incident was reported in which two girls, ages 11 and 12, were apprehended by police as they were working their way through a neighborhood selling candy to gain entrance to homes. "While one gave a sales pitch, the other would ask to use the bathroom. Once inside, the girl reportedly would take small items" (*Rocky Mountain News,* 36).

ANGER/HOSTILITY

The motives for arson among teenagers are particularly difficult to determine. Some can't offer any reasons, and in other cases there is no way to verify such reasons as may be given.

One researcher who studied arsonists concluded that they "set fires impulsively, without premeditation, in anger" (Jordan, 67). Another concluded that fire setters are frequently severely disturbed youth who may have "an inborn propensity toward antisocial behavior" (Wooden, 28). Many come from upper- or middle-class families and express a great deal of general hostility and low self-esteem (ibid.). Perhaps these fire setters have yet to outgrow the fascination with matches and fires common among smaller children. "Relatively unassertive and passive with their peers, they often find that fire satisfies their need for control and power" (Wooden, 28).

It appears that vandalism has some elements in common with arson, particularly because it may be an expression of anger or general hostility. The shattered windows of an abandoned warehouse, or the broken park bench, are all testimony to an urge to strike out, to destroy something. Peer pressure sometimes plays a key part in vandalism.

One expert has identified several categories of vandals: (1) disturbed, (2) essentially law-abiding, and (3) subcultural (Martin, 28–71). Most teenage vandals appear to belong to the third group, in which vandalism is an accepted part of daily life. Usually, though not always, this subculture is a part of the urban poverty scene. Group involvement often involves property destruction. An important subcategory of this subculture is seen in the graffiti that is usually spray-painted on any flat surface (ibid., 52–71).

COMMUNICATION

Graffiti represents a special type of vandalism. Graffiti is a form of communication for some. In the urban ghetto, it signals a gang's presence, the limits of turf. It contains warnings, messages of exploits, and challenges to rival gangs. In Chicago, gangs suddenly began a campaign of messages on train cars. "Working in spray paint and felt-tip marker, graffiti gangs 'bombed' nearly a hundred Chicago transit cars over a single weekend in June" (Uehling, 23). Graffiti also is a message to or from a special group, such as homeless youth. One investigator of this problem said: "I was surprised to find that the graffiti I encountered in the street environment so graphically displayed the antagonisms and sentiments that street youth, sometimes reluctantly, expressed verbally in interviews and informal conversations" (Luna, 74). This special type of vandalism may be the only way for some angry teens to communicate with others.

REFERENCES

Adler, Freda. "The Rise of the Female Crook," *Psychology Today* 9, no. 6 (November 1975): 42–48.

Blume, Judy. *Then Again, Maybe I Won't*. Scarsdale, NY: Bradbury Press, 1971.

Cameron, Mary Owen. *The Booster and the Snitch: Department Store Shoplifting*. New York: Free Press, 1964.

Clark, Ramsey. *Crime in America*. New York: Simon & Schuster, 1970.

Cohen, Barbara. *King of the Seventh Grade*. New York, Lothrop, Lee & Shepard, 1982.

Edwards, Loren. *Shoplifting and Shrinkage Protection*. Springfield, IL: C. C. Thomas, 1976.

Federal Bureau of Investigation. *Uniform Crime Reports of the United States*. Washington, DC: U.S. Department of Justice, 1985.

Federal Bureau of Investigation. *Uniform Crime Reports of the United States*. Washington, DC: U.S. Department of Justice, 1987.

Federal Bureau of Investigation. *Uniform Crime Reports of the United States*. Washington, DC: U.S. Department of Justice, 1988.

Flowers, Ronald. *Children and Criminality: The Child as Victim and Perpetrator*. New York: Greenwood Press, 1986.

Greenberg, Jan. *Bye, Bye, Miss American Pie*. New York: Farrar, Straus & Giroux, 1985.

Jordan, Nick. "Arson As a Weapon of Choice," *Psychology Today* 19, no. 5 (May 1985): 67.

Kraut, Robert E. "Deterrent and Definitional Influences on Shoplifting," *Social Problems* 23, no. 3 (February 1976): 363–365.

Luna, G. Cajetan. *Society* 24, no. 6 (September/October 1986): 73–78.

Martin, Albert C. *Delinquent Boys*. New York: Free Press, 1955.

McCaghy, Charles. *Crime in American Society*. New York: Macmillan, 1980.

"More Than Sticky Fingers," *Psychology Today* 22, no. 11 (November 1988): 10.

Murphy, Jim. *Death Run*. New York: Clarion, Houghton Mifflin 1982.

New York Times, 30 June 1988, A-1.

"Pilfering Urges," *Time* 116 no. 20 (27 November 1980): 94.

Reid, Sue T. *Crime and Criminology*. 5th ed. New York: Holt, Rinehart & Winston, 1988.

Rocky Mountain News, 13 April 1989: 36.

Roeback, Julian. *Criminal Typology*. Springfield, IL: Charles C. Thomas, 1966.

Sanders, William B. *Juvenile Delinquency: Causes, Patterns, and Reactions*. New York: Holt, Rinehart & Winston, 1981.

Sichor, David, and Delos Kelly. *Critical Issues in Juvenile Delinquency*. Lexington, MA: Lexington Books, 1980.

Uehling, Mark D. "Making War on Graffiti," *Newsweek* 108, no. 23 (11 August 1986): 23.

U.S. Department of Justice. *Report to the Nation on Crime and Justice*. 2d ed. Washington, DC: U.S. Government Printing Office, 1988.

U.S. Department of Justice. *Sourcebook of Criminal Justice Statistics—1988*. Washington, DC: U.S. Government Printing Office, 1989.

Wooden, Wayne S. "Arson Is Epidemic—and Spreading Like Wildfire," *Psychology Today* 19, no. 1 (January 1985): 23–28.

Resources
for Finding Out about Violence against Property

Fiction

Adler, C. S. **With Westie and the Tin Man.** New York: Macmillan, 1985. 194p.

Fifteen-year-old Greg is trying to adjust to family life after spending a year in a correctional institution for shoplifting stereo parts. Life at home means dealing with Mom's alcoholism and a live-in boyfriend who is her business partner.

Banks, Lynn Reid. **The Writing on the Wall.** New York: Harper & Row, 1981. 244p.

A teenage girl takes a trip to be with her boyfriend, and they become involved in vandalism and drug smuggling.

Blume, Judy. **Then Again, Maybe I Won't.** Scarsdale, NY: Bradbury, 1971. 164p.

What should Tony do when he discovers that his friend, Joel, has a shoplifting habit? Certainly physical pain tells him, and the doctor has told him how to deal with his nervous stomach, but Joel doesn't want to turn in a friend.

Cohen, Barbara. **King of the Seventh Grade.** New York: Lothrop, Lee & Shepard, 1982. 190p.

Vic, a 13-year-old, hates Hebrew school and reacts by shoplifting. This book tells about his attempt at shoplifting and the consequences.

Daly, Jay. **Walls.** New York: Harper & Row, 1980. 214p.

Seventeen-year-old Frankie O'Day has to deal with prankster friends, an alcoholic father, and a new romance. He helps himself relieve the pressure by becoming "The Shadow," who writes graffiti on walls.

Gauch, Pat. **Fridays.** New York: Putnam, 1979. 159p.

Corey wants to belong to the crowd, so she joins a group of girls who call themselves the Eight. Their adventures escalate into shoplifting and involvement with boys who steal a car. Corey decides that she would like a new start, without the Eight.

Grace, Fran. **Branigan's Dog.** Scarsdale, NY: Bradbury, 1981. 185p.

Casey cannot handle his feelings for and about human beings, so he relates only to his dog, Denver. Arson helps Casey enter a fantasy world where he is a hero. After Denver's death, and time in a counseling facility, Casey is able to come to terms with his human relationships.

Greenberg, Jan. **Bye, Bye, Miss American Pie.** New York: Farrar, Straus & Giroux, 1985. 150p.

Jason is a shoplifter whose actions seem designed to impress his friends. Beth sees him as a wild and exciting guy, but later discovers that his hell-raising may be aimed at getting caught.

Guy, Rosa. **New Guys around the Block.** New York: Dell/Laurel, 1987. 199p.

Imamu is spending the summer helping his mother in Harlem while a phantom burglar is terrorizing the neighborhood. Imamu suspects that the burglar may be one of his teenage friends.

Hellman, Ruth. **I Gotta Be Free.** Philadelphia: Westminster, 1979. 92p.

Jay leaves home and hooks up with a runaway brother and sister who live by robbing homes. Jay is arrested, but released, and must make some decisions about the direction of his life by standing up for himself.

Kropp, Paul. **No Way.** St. Paul, MN: EMC Publishing, 1980. 93p.

Peter is well known among his friends for his shoplifting, but after his arrest, they abandon him. He must decide whether to give in to pressure to return to that life after probation.

Lasky, Kathryn. **Prank.** New York: Macmillan, 1984. 171p.

After Birdie's brother participates in the desecration of a synagogue, his family and friends explore the reasons why. He develops a skill in working with boats, and learns how he can become real and not a prank himself.

Major, Kevin. **Far from Shore.** New York: Delacorte, 1980. 215p.

Chris's family situation and his drinking lead him to vandalism and blackouts. As a camp counselor, he is forced to recognize his own problems; perhaps his life can improve.

Morton, Jane. **Running Scared.** New York: Elsevier/Nelson, 1979. 118p.

Dave feels frustrated because of his poor self-image and problems with school. He steals a car for a joyride and is caught and detained in juvenile hall. Dave learns to change his life through his newly developed track skills.

Myers, Walter D. **Won't Know till I Get There.** New York: Viking, 1982. 176p.

Steve's family includes a 23-year-old who has been in trouble with the law. Steve commits vandalism to emulate his new brother, but he and his friends are caught. When they must work at a senior citizens home, the lives of the older people touch them.

Phipson, Joan. **Hit and Run.** New York: Atheneum, 1985. 123p.

Roland has already been in trouble for vandalizing a car. Now to escape the constable who is after him, he steals another car and strikes a baby carriage. Roland must learn to make new choices for his life and friends.

Nonfiction

Abelson, Elaine S. **When Ladies Go A-Thieving.** New York: Oxford University Press, 1989. 292p.

This book provides a comprehensive account of the underlying motives as well as the unique methods of shoplifting among females, including adolescents. Chapter 2, "The World of the Store," explains why the glamorous displays in the larger and elegant department stores are especially good targets.

Goldstein, Arnold P., et al. **Aggression Replacement Training.** Champaign, IL: Research Press, 1987. 376p.

This book tells how to develop anger control techniques and moral reasoning in teenagers. The procedures described are very useful in combating crimes, especially those directed at impersonal targets.

Sanders, William B. **Juvenile Delinquency: Causes, Patterns, and Reactions.** New York: Holt, Rinehart & Winston. 1976. 238p.

Information about teenage criminal activity is provided, and Chapter 7, "Juvenile Property Crimes," is an especially relevant part of the book.

Sklar, Stanley L. **Shoplifting: What You Need To Know about the Law.** New York: Fairchild Publications, 1982. 229p.

Consumers, including teens, are sometimes detained for shoplifting, and this information could be especially useful. A valuable feature of this book is the Statutory Appendix, which contains laws concerning shoplifting in each of the 50 states. The book provides information on how the law deals with those who are accused, whether rightfully or not.

Taylor, Lawrence. **Born to Crime.** Westport, CT: Greenwood Press, 1984. 179p.

This book examines the genetic aspects of criminal behavior. It analyzes carefully the nature-vs.-nurture issue in regard to antisocial behavior, as well as many other scientific angles.

Wooden, Wayne S., and Martha Lou Berkey. **Children and Arson: America's Middle Class Nightmare.** New York: Plenum Press, 1984. 267p.

Appropriately subtitled, this book is a comprehensive description of juvenile fire setters. The strength of this book is its analysis of motives and some of the emotional underpinnings of young arsonists.

Yablonsky, Lewis, and Martin R. Haskell. **Juvenile Delinquency.** 4th ed. New York: Harper & Row, 1988. 510p.

Coverage of a wide range of criminal activity among teenagers. Chapter 7, "The Juvenile in a Violent Society," is a detailed explanation of violent behavior. The section on senseless violence, pages 225 to 231, is especially good.

Nonprint Materials

The Boy Who Liked Deer

Type: 16mm film; color
Length: 18 min.
Cost: Rental $13
Distributor: Learning Corporation of America
1350 Avenue of the Americas
New York, NY 10019
Date: 1975

The story of several boys whose anger leads them to vandalism that unintentionally results in the death of a deer. This brutal killing is followed by a streak of vandalism in the English teacher's classroom. The inner pain felt by the boy who discovers the deer is emphasized.

A Circle in the Fire

Type: 16mm film; color
Length: 50 min.

Cost: Rental $23
Distributor: Perspective Films
65 East South Water Street
Chicago, IL 60601
Date: 1975

A film adaptation of Flannery O'Connor's short story about the destruction and disruption caused by three disturbed teenage boys. The story progresses from the idyllic atmosphere of a southern farm to a blazing climax. The film starts slowly and moves steadily to a dramatic finale.

Graffiti
Type: 3/4″ video
Length: 15 min.
Rent from: University of Connecticut
Center for Institutional Media & Technology
Storrs, CT 06268
(203) 486-2530 (for cost and arrangements)
Date: 1974

This film makes the point that graffiti is a form of environmental pollution, and the cost of removing it runs into millions of dollars each year. Though some may see it as an art form, it is a major problem in most large urban centers.

Greenhouse
Type: 16mm film, video; color
Length: 11 min.
Cost: Purchase $215 (film), $150 (1/2″ video)
Distributor: Barr Films
P.O. Box 7878
Irwindale, CA 91706
(800) 234-7878
Rent from: Boston University
Krasker Memorial Film Library
565 Commonwealth Avenue
Boston, MA 02215
(617) 353-3272 (for cost and arrangements)
Date: 1973

A boy's vandalism and his subsequent efforts to help the owner of a greenhouse repair the damage lead to a new friendship. The old man who is the greenhouse owner provides for the boy's learning and some new values to live by.

Kids and Guns

Type: Various video; color
Length: 28 min.
Cost: Rental $75 (plus shipping); purchase $149 (VHS or Beta), $199 U-Matic
Distributor: Films for the Humanities & Sciences
P.O. Box 2053
Princeton, NJ 08543
(800) 257-5126
Date: 1988

The American gun culture is described in this film, as well as its impact on youth. How the mania for guns developed in this country is a focal point. Some ideas are offered about how we might stem the increase in gun accidents and violence.

The Kids Case against Vandalism

Type: 16mm film, VHS (1/2″); color
Length: 12 min.
Cost: Rental (call distributor); purchase $250 (film), $150 (VHS)
Distributor: Landmark Films
3450 Slade Run Drive
Falls Church, VA 22042
(800) 342-4336
Date: 1982

This film presents a reenactment of vandalism involving three boys who have broken into a place of business and caused extensive damage. The personalities of the boys, their motives, and a trial provide viewers with additional insight into the crime of vandalism.

Portrait of a Teenage Shoplifter

Type: Various video, color
Length: 47 min.

Cost: Rental not available from distributor, though preview can be arranged; purchase $49.50; add $50 for 3/4″ format
Distributor: Ambrose Video Publishing, Inc.
381 Park Avenue South, Suite 1601
New York, NY 10016
(800) 526-4663
(212) 696-4545 in New York
Date: 1984

This film is taken from the ABC Afterschool Special television series, which won Peabody and Emmy awards. A high school cheerleader steals for no apparent reason other than fun, but when caught, she faces severe consequences.

Portrait of a Vandal
Type: Various video; color
Length: 13 min.
Rent from: Kent State University
Audio Visual Services
330 Library Building
Kent, OH 44242
(216) 672-3456 (for cost and arrangements)
Date: 1979

Because Debbie appeared to some of the guys to be stuck-up, they waited until she and her family were gone, then "trashed" the house. This case is based on actual police records and shows how mindless and stupid, yet vicious, vandalism really is.

School Vandalism
Type: 16mm film, various video; color
Length: 10 min.
Cost: Purchase $180 (film), $110 (video)
Distributor: AIMS Media
6901 Woodley Avenue
Van Nuys, CA 91406

Rent from: Idaho State University
Audio-Visual Services
Box 8064—Campus
Pocatello, Idaho 83209
(208) 236-3212 (for cost and arrangements)
Date: 1972

Four boys who are bored decide to break into their school, where they accidentally set fire to the cafeteria. The film opens with them sitting in a jail cell waiting for their parents. Then the entire episode is recreated for viewers.

Shoplifting
Type: 16mm film; color
Length: 21 min.
Cost: Purchase $420 (film), video $110
Distributor: AIMS Media
6901 Woodley Avenue
Van Nuys, CA 91406
Rent from: University of Maine
Instructional Systems Center
12 Shibles Hall
Orono, ME 04469
(207) 581-2510 (for cost and arrangements)
Date: 1973

The motives and dynamics surrounding shoplifting are discussed by psychiatrists, retail security officers, and police. The effects of this crime are examined in several ways, including a depiction of a woman who was caught shoplifting and the resulting reaction of her husband. There is a reasonably good emphasis on the economic impact of shoplifting, particularly on the fact that the cost of the crime must be passed along to all consumers.

Shoplifting Is Stealing
Type: 16mm film, various video; color
Length: 16 min.
Cost: Purchase $320 (film), $110 (video)

Distributor: AIMS Media
6901 Woodley Avenue
Van Nuys, CA 91406
Rent from: Boston University
Krasker Memorial Film Library
565 Commonwealth Avenue
Boston, MA 02215
(612) 353-3272 (for cost and arrangements)
Date: 1975

Teenagers are reminded that shoplifting is a crime. This film shows several case histories of shoplifting and explores the means that retailers use to protect themselves.

Shoplifting: It's a Crime
Type: 16mm film, various video; color
Length: 12 min.
Cost: Rental $20 (three days); purchase $235 (film), $89 (video)
Distributor: Filmfair Communications
10900 Ventura Boulevard
P.O. Box 1728
Studio City, CA 91604
Rent from: Boston University
Krasker Memorial Film Library
565 Commonwealth Avenue
Boston, MA 02215
(612) 353-3272 (for cost and arrangements)
Date: 1975

This film emphasizes the long-range consequences of shoplifting, as well as the more immediate process of detention and jail. Kathy is caught stealing cosmetics and jewelry from a department store, and we see her being apprehended by the police, taken to the station for fingerprinting and detention, and finally placed in a jail cell.

Shoplifting: Preventing the Crime
Type: 16mm film, various video
Length: 23 min.

Cost: Rental $75 (one week); purchase $480 (film), $360 (video)
Distributor: AIMS Media
6901 Woodley Avenue
Van Nuys, CA 91406
(800) 367-2467
Date: 1986

Shows the details of various methods of shoplifting.

Shoplifting: Sharon's Story

Type: Various video; color
Length: 26 min.
Distributor: Learning Corporation of America
1350 Avenue of the Americas
New York, NY 10019
Rent from: Boston University
Krasker Memorial Film Library
565 Commonwealth Ave.
Boston, MA 02215
(612) 353-3272 (for cost and arrangements)
Date: 1978

Sharon is apprehended while shoplifting, and viewers follow the entire detention/justice process. Sharon is searched, interrogated, and detailed, then goes to court with her family.

Shoplifting: You Pay for It

Type: 16mm film, various video; color
Length: 16 min.
Cost: Rental $75 (three days); purchase $365 (film), $320 (video)
Distributor: Coronet/MTI Film & Video
108 Wilmot Road
Deerfield, IL 60015
(800) 621-2131; (708) 940-3600
(call collect from Illinois and Alaska)
Date: 1982

Points out the economic cost of shoplifting to consumers. When merchandise is stolen, overhead costs rise and the prices must then be raised for all other shoppers. The emphasis in this film is on the cost of the crime of shoplifting to society rather than the process of detention and justice.

Short Sharp Shock

Type: Video
Length: 27 min.
Cost: Rental (contact distributor)
Distributor: Concord Films Council
201 Felixstowe Road
Ipswich, Suffolk IP3 9BJ
England
Date: 1983

Teenagers describe how it feels to be in police detention. Despite the sensitivity of police, those who are detained feel scorned and rejected.

So I Took It

Type: 16mm film, various video; color
Length: 10 min.
Cost: Purchase $265 (film), $240 (video)
Distributor: Coronet/MTI Film & Video
108 Wilmot Road
Deerfield, IL 60015
(800) 621-2131; (708) 940-3600
(call collect from Illinois and Alaska)
Date: 1975

Shoplifting occurs as a result of peer pressure. The teenager seldom has a pressing economic need for the stolen items, but rather tries it as a "stunt" to talk about later.

Teenage Shoplifting

Type: Various video; color
Length: 10 min.
Cost: Purchase $99 (film), $59 (1/2"), $69 (3/4")
Distributor: CRM
2233 Farraday Avenue
Carlsbad, CA 92008
(800) 421-0833

Rent from: Kent State University
Audio Visual Services
330 Library Building
Kent, OH 44242
(216) 672-3456 (for cost and arrangements)
Date: 1981

A positive program in Michigan aimed at preventing teenage shoplifting. Here we see both parents and members of the community working cooperatively to reduce shoplifting incidents. Some shoplifters who were caught are shown in interviews, and their motives are examined.

The Vandalism Film
Type: 1/2″ video
Length: 12 min.
Cost: Purchase $49
Distributor: Barr Films
Box 5667
Pasadena, CA 91107
(800) 234-7878
Date: 1976

Why do people vandalize? This film offers an in-depth look at the motives of vandals and suggested programs of remedy.

Vandalism: Mark of Immaturity
Type: 16mm film, various video; color
Length: 12 min.
Cost: Purchase $250 (film), $125 (video)
Distributor: AIMS Media
6901 Woodley Avenue
Van Nuys, CA 91406
(800) 367-2467
Rent from: University of Wisconsin-Madison
Bureau of A-V Instruction
1327 University Avenue
P.O. Box 2093
Madison, WI 53701-2093
(800) 367-2467 (for cost and arrangements)
Date: 1977

This film explores the reasons that children vandalize, emphasizing that for most it is a thrill-seeking or fun activity. Viewers are also reminded that others are usually hurt.

Vandalism: What and Why

Type: Various video; color
Length: 12 min.
Rent from: Indiana University
Audio Visual Center
Bloomington, IN 47405
(800) 552-8620 (for cost and arrangements)
Date: 1973

This film features a group discussion of the psychological bases of this behavior. In addition to an understanding of the enormity of the problem, viewers are provided with suggestions for alleviating or reducing it.

Vandalism—Why?

Type: 3/4″ video; color
Length: 11 min.
Distributor: Alfred Higgins Productions, Inc.
9100 Sunset Boulevard
Los Angeles, CA 90069
Rent from: University of Missouri
Academic Support Center
505 E. Stewart Road
Columbia, MO 65211
(314) 882-3601 (for cost and arrangements)
Date: 1972

Examines in depth the motives for vandalism. Although absolute answers aren't available, viewers are given numerous suggestions for how constructive action can be used when energies are diverted from destructive acts.

The Vandals

Type: Various video; color
Length: 25 min.
Rent from: Portland State University
Film Library
Portland, OR 97270
(503) 229-8490 (for cost and arrangements)

Date: 1972

An examination of the social causes of vandalism is narrated by Harry Reasoner, who explores the motives behind all types of vandalism, from the inconsequential to the malicious crime of arson.

Violence and Vandalism

Type: 16mm film; color
Length: 15 min.
Rent from: University of Arizona
Film Library
Audio Visual Building
Tucson, AZ 85721
(602) 621-3782 (for cost and arrangements)
Date: 1971

Hugh O'Brien hosts this film, which attempts to answer the basic questions concerning vandalism and violence. Both inner-city and suburban kids attempt to provide the answers as well as do expert sociologists.

CHAPTER 7

The Juvenile Justice System

When the cops stop, I figure they're going to hassle my uncle. Instead, they come straight to me and ask me if my name is mine—which I think is sort of silly.

"Yeah, I'm me."

"Better come with us, Nicky. The detectives down at the station want to talk to you."

"But my uncle . . . "

"We'll see that he gets home." The cop takes hold of my arm lightly.

I look up at him. "If you're a real cop, you'll read me my rights—like they do on TV."

"Yeah, kid, when we get to the car."

Larry Bograd, *Bad Apple*
(New York: Farrar, Straus & Giroux, 1982), 149.

In Larry Bograd's novel, *Bad Apple,* 15-year-old Nicky allows his buddy Prune to persuade him to attempt a robbery. Nothing goes right, Nicky ends up assaulting an elderly couple, and within hours the police are at his house to arrest him.

Statistics on Juvenile Justice

- Sixteen percent of all arrests for crime in the United States in 1988 involved people 18 years old and under (FBI, 8)

- The mayor of New York City recently called for tougher treatment of juveniles, including fingerprinting of youths age 13 to 15 when arrested for a felony (Barron, L-27)
- Of all juveniles apprehended for crime by the police, 83 percent are not detained. Of those held in detention, 50 percent are released to go home with their parents within four hours (Bortner, 29)
- Youths who are referred to juvenile court for a second time before age 16 are likely to become chronic offenders (Speirs, 3)

Definitions

Felony. A crime of a serious nature that may be punishable by fine or imprisonment, for example, the theft of a new Mercedes.

Misdemeanor. A less serious crime, usually not punishable by imprisonment, for example, shoplifting a 50-cent ballpoint pen.

Probation. Releasing the defendant into the community under the supervision of an officer of the court.

Juvenile. Under the federal Juvenile Delinquency Act, a juvenile is a person who has not attained his 18th birthday (Black, 779).

Juveniles or Adults?

Historically there has developed in this country a separate process for handling lawbreakers when they are children. The law specifies different treatment for juveniles than for adults. The question of when one is a juvenile, and when not, varies somewhat from state to state and according to the nature of the illegal act. As the following chart indicates, people who are 18 or older are generally treated as adults.

Table 7.1 Maximum Age of Initial Juvenile Court Delinquency Jurisdiction

State	*Age*
Wyoming	19 years
Alabama, Alaska, Arizona, Arkansas, California, Colorado, Delaware, District of Columbia, Florida, Hawaii, Idaho, Indiana, Iowa, Kansas, Kentucky, Maine, Maryland, Minnesota, Mississippi, Montana, Nebraska, Nevada, New Hampshire, New Jersey, New Mexico, North Dakota, Ohio, Oklahoma, Oregon, Pennsylvania, Rhode Island, South Dakota, Tennessee, Utah, Virginia, Washington, West Virginia, Wisconsin	18 years
Georgia, Illinois, Louisiana, Massachusetts, Michigan, Missouri, South Carolina, Texas	17 years
Connecticut, New York, North Carolina, Vermont	16 years

Source: Reproduced, by permission of McGraw-Hill, from H. Ted Rubin, *Juvenile Justice: Policy, Practice, and Law*, 2d ed. (New York: Random House, 1985), 8.

In the case of very serious offenses, some state laws permit juveniles to be tried as adults. In East St. Louis, two teenage boys pleaded guilty to armed violence, and the younger—a 14-year-old with a lengthy record—was sentenced to 120 years in prison (Gest, 50). A Denver prosecutor recently announced that 17-year-old Robby Valenzuela will be tried as an adult in a murder case ("Teen To Be Tried as Adult in Stabbing Death," 55). But for lesser crimes, prosecutors and courts have treated juveniles in a separate manner, usually referred to as the juvenile justice system.

STATUS OFFENSES

There has also long been recognized a category of offenses that pertain only to children. These are usually referred to as status offenses, because they involve the status of nonadults. "Status offenders are children who ditch school, flout parental rules, run away, drink alcoholic beverages, and do other things that get them in trouble. Adults who perform these same acts or who have similar problems are not subject to court sanctions" (Rubin, *Juvenile Justice,* 51). Perhaps the most agonizing of these status offenses is that of running away. Legal experts have long questioned the appropriateness of traditional punishment by detention for this category of offense (ibid.).

THE JUVENILE JUSTICE PHILOSOPHY

The trend toward a juvenile justice system that is different from that for adults has been apparent for many years. Corrections experts and jurists have long recognized the need to help young lawbreakers change their behavior and become productive citizens. This is in marked contrast to the idea that society should merely put them in jail as punishment or deterrent. It represents a belief that children are less responsible for their behavior than are adults. Psychologists and sociologists believe that when children break the law they are rejecting or disregarding the norms established by the adult society. However, any society must maintain norms of suitable and allowable behavior in order to protect the majority of citizens. So, the idea has long existed that children who violate the law are the product of a society (or at least a family) that has failed them. Further, they see the family as dysfunctional, and the youthful offender is therefore seen as a victim.

Another aspect of this view of the juvenile is that children have not yet fully developed their reasoning abilities. They may not be able to fully anticipate the consequences of their actions or to measure the effects of their antisocial behavior on others. Obviously this approach to juvenile justice has sparked debate and controversy. Some persons, including many teens, see this approach as softheaded and ineffective. Some advocates of tough "law and order" believe that only harsh penalties along the lines of adult sentences can deter youth from continuing criminal behavior.

Nonetheless, in most states today, juveniles are treated differently from adult lawbreakers, and separate procedures are usually followed by both the police and the courts. This philosophy of treating juveniles less harshly than adult offenders may well change as public opinion shifts. Many adults believe that stiffer penalties and longer jail sentences are the answer to juvenile crime. There is some evidence that teenage offenders themselves see the present system as ineffective. When a 15-year-old who was accused of raping a woman at knifepoint was arrested, he sneered at the police, "What you gonna do to me? Send me to the youth hall. I'll be out in a few hours!" (Warner, 19). In any case, the philosophy of how to treat juvenile offenders varies from one community to another and may well be changing to meet public pressure.

What Ha[illegible]nders

What happens when [illegible] crime depends, of course, upo[illegible]d the special circumstances su[illegible]e, the shoplifter who is caught red-h[illegible]ers or other employees of a retail store. W[illegible]e summoned to the scene, the manager of the st[illegible]ecide to press charges, or he may not. Sometimes a stern re[illegible]mand by the store manager, and the youth's offer to pay for or replace the stolen item, is regarded as sufficient. The juvenile shoplifter might then be released to parents with instructions by the policeman to "straighten out" or to "stay out of trouble." If the value of the stolen item is significant, or the store manager insistent, the offender may be taken to the police station for intake. The term *intake* refers to what takes place after apprehension by the police.

INTAKE

This is the initial process, in which the court tries to determine the facts of the case and what type of referral might be most appropriate. The exact nature of the process varies from state to state, and even from court to court within a state. The court must decide if it has jurisdiction in a particular case, and, if so, if the case should proceed through the full court process. Depending upon the facts that are available, the court has numerous options at this point. One of these options, of course, is filing, which is the making of a formal charge. Another is dismissal. Dismissal might occur because the evidence is weak, or it might be delayed while the violator performs some specified work or community service. Another option for the court is referral. This referral could be a form of diversion—some type of work or behavior that the defendant voluntarily agrees to. The referral could involve the local school, the parents, the police, or any community agency with the potential for education and correction.

DIVERSION

Some offenders are offered opportunity to make restitution for their offense, or to pursue an option other than a jail sentence. These options are usually referred to as diversion, and can occur at various steps throughout the process of arrest and court activity. For example, if caught in the act of spray-painting graffiti on walls or sidewalks, the offenders might be detained only long enough to be lectured about lawbreaking behavior, then released. But when the offenders are offered the opportunity of cleaning up their graffiti (and perhaps some left by other vandals) instead of a trip to the police station for booking, diversion has occurred. Diversionary options might occur later as the case reaches the public prosecutor. In the course of court proceedings, juvenile judges sometimes propose diversions such as restitution, remedial education, or community service. This is discussed in greater detail in a later section.

TYPICAL COURT PROCEEDING

The following excerpted dialogue is from *Behind the Black Robes*, by H. Ted Rubin. Judge Rubin has constructed what he regards as a typical day in juvenile court, and it provides a good sample of what juvenile offenders might expect.

> The judge calls the first case, a burglary, and looks across at the thirteen-year-old boy he has not seen before. . . . The juvenile courtroom is often less austere

than other courtroom structures. The judge's bench, frequently, is just one-step-up, a deliberate design to reduce formality and facilitate communication between the judge, the child, and other participants. Judicial communication with youngsters can be an art form, counted by many judges as an important skill, indeed, a necessity. Others, less artful, use the law and procedural formalities as their canvas.

You are James Bellamy—do people call you Jim? . . .

This is the communicative judge beginning his work. He has quickly sensed that the child's age, offense, and first offense status do not mandate strict formalities. The judge can begin to balance his awesome powers with a first attempt to show interest.

Un-huh. . . .

Some judges prefer a "yes, your honor" response, but this will bother another judge who thinks the polite response has been rehearsed. Many judges are pleased to obtain even a minimal verbal answer, one that does not require the judge to point out that the hearing is being recorded and the record does not pick up nods of the head.

And you are Jim's mother? . . .

A parent is required to attend juvenile court hearings. Several reasons support this requirement. The child is a minor and, despite a law violation charge, is a legal non-person. Accordingly, a parent's consent to how the child exercises his legal rights in a court proceeding is a necessity. The second reason is the public policy position that parents should be involved when their child is involved with the court (11–12).

Following this somewhat informal opening by the judge, other court officials, particularly a probation officer, may be introduced to both the juvenile and the parents. Then the judge will likely advise the juvenile of his/her legal rights. The 1967 U.S. Supreme Court ruling in the Gault case established that children are entitled to due process. Therefore, the court will probably advise the juvenile of the right to a lawyer's assistance. For a more complete account of a typical juvenile court proceeding, you are urged to read Chapter 1 of *Behind the Black Robes*.

DISPOSITION OF CASES

Judges who handle juvenile offenders are generally looking for a way to make the punishment in some way correctional or educational. Probation officers are often involved in the deliberations about the offender, and of course every attempt is made to bring the parents into the picture. Sometimes, when the parents are unavailable or the family is dysfunctional, the court may assign the offender to a foster home and ask that reports come back to the court concerning the offender's future behavior.

The concept of restitution is important to most of today's juvenile court proceedings. The restitution may be some form of reimbursement to the victim of the crime, or a general form of recompense such as community service. Juvenile courts try very hard to find alternatives along these lines, recognizing that serving time in a jail or detention facility is a last resort. However sensitive courts may be in this respect, judges are not patsies, and repeat offenders, particularly those who show no remorse, may find themselves doing time in detention.

The unanswered question for society and juvenile justice experts concerning youthful offenders is "what works best?" The research evidence is not that clear nor that consistent. Most experts agree that the emphasis should remain on correction rather than on punishment alone. In the meantime, there is increasing public pressure upon the courts and law enforcement officials to "keep the thugs off the streets" and protect all citizens. Teens who run afoul of the law can expect the courts to arrange for some kind of correction for the first-time offender, particularly if the crime is not very serious. Repeat offenders and those who commit crimes of violence may well feel the full force of a society that has decided it won't take it anymore.

REFERENCES

Barron, James. "Koch Urges Tough Action on Juvenile Offenders," *New York Times*, 14 May 1989: L-27.

Black, Henry G. *Black's Law Dictionary*. 5th ed. St. Paul: West Publishing, 1979.

Bograd, Larry. *Bad Apple*. New York: Farrar, Straus & Giroux, 1982.

Bortner, M. A. *Inside a Juvenile Court*. New York: New York University Press, 1982.

Federal Bureau of Investigation, U.S. Department of Justice, *Uniform Crime Reports for the United States*. Washington, DC: U.S. Government Printing Office, 1988.

Gest, Ted. "Kids, Crime, and Punishment," *U.S. News and World Report* 130, no. 8 (24 August 1987): 50–51.

Rubin, H. Ted. *Behind the Black Robes*. Beverly Hills, CA: Sage, 1985.

———. *Juvenile Justice: Policy, Practice, and Law*. 2d ed. New York: Random House, 1985.

Speirs, Verne L. "Study Sheds New Light on Court Careers of Juvenile Offenders," in *OJJDP Update on Research*. Washington, DC: Office of Juvenile Justice and Delinquency Prevention, Department of Justice, August 1988.

"Teen To Be Tried as Adult in Stabbing Death," *Rocky Mountain News*, 16 December 1988: 55.

Warner, Edwin. "The Youth Crime Plague," *Time* 100, no. 2 (11 July 1977): 18–28.

Resources
for Finding Out about the Juvenile Justice System

Fiction

Bograd, Larry. **Bad Apple.** New York: Farrar, Straus & Giroux, 1982. 152p.

Fifteen-year-old Nicky and his pal Prune have never been very successful at anything, and when they undertake armed robbery, everything goes wrong. This book includes an excellent and exciting account of Nicky's arrest and how the juvenile justice system handled him.

Bridgers, Sue Ellen. **Permanent Connections.** New York: Harper & Row, 1987. 320p.

Rob's behavior has been annoying to his parents—in many ways he's alienated. Spending his summer in North Carolina with relatives seems to be the perfect solution, especially since his uncle is injured and some extra help would be welcome. All goes well until Rob goes on a binge with pot and booze and ends up in a wreck. A sheriff's deputy arrests him and he is prosecuted. Rob's day in court is the climax of this book and will engage teens in an experience with the justice system.

Butterworth, W. E. **The Narc.** New York: Four Winds, 1972. 187p.

Young Dan Morton, who recently graduated from the police academy, is assigned to undercover work at the local high school. Here's an inside look at police work—the intense contacts with high school students—and the tension. This book tells how Dan operates as the key link in law enforcement with teenagers and the justice system.

Samuels, Gertrude. **Run, Shelley, Run!** New York: Signet, New American Library, 1974. 192p.

Cottage C at the state training schools for girls is grim and destructive, but Shelley has had nothing but trouble from mom, stepdad, and almost everybody else. This is a gripping story of a teenage girl and her struggle against society and the justice system.

Nonfiction

Bortner, M. A. **Inside a Juvenile Court: The Tarnished Ideal of Individualized Justice.** New York: New York University Press, 1982. 328p.

One of the basic tenets of modern-day juvenile justice philosophy is that the needs of the juvenile should be emphasized over traditional punishment. Bortner's book looks closely at this and other underlying principles of juvenile justice in the United States. Though the book seems aimed toward adult readers, the reading level is within reach of many teens. The entire issue of the system's fairness to racial minorities is very well analyzed in Chapter 9, "The Influence of Social Characteristics: The Issue of Race."

Davidson, William S., Robin Redner, Richard Amdur, and Christina M. Mitchell. **Alternative Treatments for Troubled Youth.** New York: Plenum Press, 1990. 292p.

By most measures, the traditional juvenile justice system has not served either society or young persons well. These authors provide information about diversion programs and what makes them successful.

Dolan, Edward F. **Protect Your Legal Rights.** New York: Julian Messner, 1983. 128p.

This timely book focuses on legal rights of juveniles. For example, the author asks: "As a runaway, do I have any legal rights at all?", then provides the answer. This question and answer format is effective and easily understood. The topics range from contracts to involvement with the police and appearances in court.

Epstein, Sam, and Beryl Epstein. **Kids in Court.** New York: Four Winds, 1982. 233p.

The American Civil Liberties Union has been involved in many court battles on behalf of young people. Eleven of these cases are presented here in a style that is highly readable. The legal rights of teenagers who are in trouble with the law are made quite clear in each case. One of the most timely of these is a case involving invasion of privacy when a school psychologist surveys potential drug abusers among junior high students.

Kramer, Rita. **At a Tender Age: Violent Youth and Juvenile Justice.** New York: Henry Holt, 1988. 309p.

Extensive research in New York City's juvenile justice system led Kramer to this extensive report on the system's shortcomings. Her view is that the system was set up to deal with truants and kids who commit petty crimes, rather than the murderers and robbers who are now its clientele.

Olney, Ross R. **Up against the Law: Your Legal Rights as a Minor.** New York: Dutton, 1985. 132p.

Teenagers often need to know exactly where they stand in reference to the law. Olney's book addresses numerous areas of legal concern and focuses on the rights of nonadults.

Rubin, H. Ted. **Behind the Black Robes.** Beverly Hills, CA: Sage, 1985. 248p.

An excellent introduction to the complexities and issues involved in juvenile court. The beginning chapter offers a script-like interchange between judge and offender, just as it might occur in the court setting. The roles of social worker, probation officer, lawyers, and parents are illustrated in this highly effective format. This is "must" reading for the teen or adult who wants to know how juvenile courts really operate.

———. **Juvenile Justice: Policy, Practice, and Law.** 2d ed. New York: Random House, 1985. 419p.

This book offers the most up-to-date and comprehensive account of how juvenile justice operates across the United States.

Swiger, Elinor P. **The Law and You: A Handbook for Young People.** Indianapolis: Bobbs-Merrill, 1973. 214p.

There are many situations in which young people potentially may be victims or offenders, and this book discusses a number of them. Chapter 6, titled "Are There Special Courts for Young People?", is especially relevant. A glossary of terms at the end of the book translates legal and court jargon into simple language.

Nonprint Materials

But It's Not My Fault

Type: 16mm film, video; color
Length: 46 min.
Cost: Rental $75; purchase $750
Distributor: Texture Films, Inc.
1600 Broadway
New York, NY 10019
Date: 1982

A 16-year-old has become a delinquent in response to peer pressure. However, he decides to give up delinquency after spending a weekend in jail with some more experienced criminals.

Crime, Punishment . . . and Kids

Type: 1/2″ video; color
Length: 50 min.
Cost: Purchase $79
Distributor: Films Inc.
5547 N. Ravenswood Avenue
Chicago, IL 60640-1199
(800) 323-4222, ext. 43
Date: 1987

This video, from an NBC special dealing with juvenile crime and justice, explores the balance between rehabilitation of the offender and the public's right to protection from the offender on the street.

The Criminal Justice System

Type: 16mm film, various video; color
Length: 23 min.
Cost: Rental $75 (three days)(plus shipping and handling); purchase $475 (film), $420 (video)
Distributor: Coronet/MTI Film & Video
108 Wilmot Road
Deerfield, IL 60015
(800) 621-2131; (708) 940-3600
(call collect from Illinois and Alaska)
Date: 1982

Part 2 of a two-part documentary that examines crime in America and how the justice system functions, this film shows how the probation system often contributes to the problem rather than solving it.

Delinquency: The Chronic Offender

Type: 3/4″ video
Length: 28 min.
Rent from: University of Iowa
Audio-Visual Center
Seashore Hall
Iowa City, IA 52242
(319) 353-5885 (for cost and arrangements)
Date: 1977

In a case study of "Shotgun Joe," a sociopath, we look inside his prison world and examine his interrelationships with other people. The film emphasizes the shortcomings of our present juvenile corrections programs and challenges the viewers to think of better ways of dealing with juvenile repeat offenders.

Delinquency: Prevention and Treatment

Type: 3/4″ video
Length: 28 min.

Rent from: Syracuse University
Film Rental Center
1455 E. Colvin Street
Syracuse, NY 13210
(315) 479-6631 (for cost and arrangements)
Date: 1977

Three contrasting types of delinquency prevention and treatment programs are explored in this video. One is a diversion program in Berkeley that uses police officers as positive role models. The second is the Positive Peer Culture program at Red Wing, Minnesota. The third is a community-based group home in Pittsburgh in which love and strict discipline are the primary techniques.

Juvenile Delinquency—It's Up to You (Law and Juvenile Series)
Type: 16mm film, various video; color
Length: 19 min.
Cost: Purchase $425 (film), $300 (1/2″ video), $330 (3/4″ video)
Distributor: Barr Films
Box 7878
Irwindale, CA 91706-7878
Rent from: Kent State University
Audio Visual Services
330 Library Building
Kent, Ohio 44242
(216) 672-3456 (for cost and arrangements)
Date: 1978

Teenage lawbreakers discuss their experiences, including their feelings and motivations. Their comments provide considerable insight into juvenile law and some of the basic differences between the processes of the adult and the juvenile justice systems.

Juvenile Law
Type: 16mm film, color
Length: 23 min.

Rent from: Syracuse University
Film Rental Center
1455 E. Colvin Street
Syracuse, NY 13210
(315) 479-6631 (for cost and arrangements)
Date: 1973

Through the experiences of two brothers ages 15 and 18, the film shows how juvenile offenders are treated differently from adult criminals. The brothers are arrested for a crime in which both participate. One, however, is processed under adult statutes, and the other—in stark contrast—as a juvenile.

Law and Youth

Type: Filmstrip, sound, color
Length: Approximately 16 min. each (two filmstrips)
Cost: Rental (free 30-day preview available); purchase $77 for both
Distributor: Encyclopedia Britannica Educational Corp.
310 S. Michigan Avenue
Chicago, IL 60604
(800) 554-9862
Date: 1981

Two companion sound filmstrips about teens and the law. "The Juvenile: Adult or Minor" concerns the problem of dealing with an offender as either an adult or a child. "Juvenile Justice: Rights and Responsibilities" deals with the balance that the juvenile justice system must seek between protecting citizens and locking up offenders.

Project Aware

Type: Various video, color
Length: 27 min.
Rent from: Washington State University
Instructional Media Services
Pullman, WA 99164
(509) 335-5618 (for cost and arrangements)
Date: 1977

Young people are told directly by an ex-convict what it's really like in prison. He points out his early mistakes and the long path he took that eventually led him to prison and the horrors he experienced there.

The Reluctant Delinquent

Type: Various video
Length: 24 min.
Rent from: Indiana University
Audio Visual Center
Bloomington, IN 47405
(812) 335-2103 (for cost and arrangements)
Date: 1977

This video takes a look at Robbie, a 17-year-old in maximum security, and the delinquency that landed him there. Robbie's dyslexia, which created problems for him at school, was a major factor. However, as is shown, delinquency is much more than a learning disability.

Short Sharp Shock

Type: Video
Length: 27 min.
Cost: Contact distributor for latest price
Distributor: Concord Films Council
201 Felixstowe Road
Ipswich, Suffolk IP3 9BJ
England
Date: 1983

Teenagers describe how it feels to be in police detention. Despite the sensitivity of police, those who are detained feel scorned and rejected.

The Society

Type: 16mm film, various video
Length: 29 min.
Cost: Rental $75 (three days)(plus shipping and handling); purchase $520 (film), $470 video
Distributor: Coronet/MTI Film & Video
108 Wilmot Road
Deerfield, IL 60015
(800) 621-2131; (708) 940-3600
(call collect from Illinois and Alaska)
Date: 1982

Deals with the societal factors that contribute to crime in the United States. This is part one of a two-part documentary dealing with crime and the justice system.

Violent Youth: The Un-Met Challenge
Type: Various video; color
Length: 23 min.
Rent from: Indiana University
Audio Visual Center
Bloomington, IN 47405
(800) 552-8620 (for cost and arrangements)
Date: 1979

John and William, who are in a juvenile detention home, are interviewed about their past behavior and aspirations for the future. The film deals with violent juvenile crime and shows the viewer the processes of the juvenile justice system from detention to punishment.

Who Wants To Be a Hero?
Type: 16mm film, various video; color
Length: 28 min.
Cost: Rental $75 (three days)(plus shipping and handling); purchase $500 (film), $310 (video)
Distributor: Coronet/MTI Film & Video
108 Wilmot Road
Deerfield, IL 60015
(800) 621-2131; (708) 940-3600
(call collect from Illinois and Alaska)
Date: 1981

Witnesses are crucial to the trials of offenders under our justice system. Jason agrees to testify against another guy in a trial for assault, but after his family is terrorized, he begins to have second thoughts.

Organizations

Correctional Education Association
8025 Laurel Lakes Court
Laurel, MD 20707
(301) 490-1440

Executive Director: Dr. Stephen Steurer

The association attempts to increase the effectiveness of all who work in corrections settings, both adult and juvenile.

PUBLICATION: *Journal of Correctional Education*, quarterly.

Legal Services for Children
1254 Market Street
San Francisco, CA 94102
(415) 863-3762
Executive Director: Carole Brill

The group provides legal services for children and teenagers in the San Francisco Bay area.

PUBLICATION: *Parents' Guide to Special Education.*

National Association of Counsel for Children
1205 Oneida Street
Denver, CO 80220
(303) 321-3963
Executive Director: Donald Bross

The association works to develop and improve laws concerning juveniles and promotes legal training and better representation in court for children.

PUBLICATION: *The Guardian*, quarterly.

National Center for Juvenile Justice
701 Forbes Avenue
Pittsburgh, PA 15219
(412) 227-6950
Director: Hunter Hurst

Encourages progressive administration of justice to juveniles and disseminates juvenile court statistics.

PUBLICATION: *Today's Delinquent*, annual.

National Council on Crime and Delinquency
77 Maiden Lane (4th Floor)
San Francisco, CA 94180
(415) 956-5651
President: Barry Krisberg

Provides information and training for those interested in juvenile and family courts or prevention and control of crime and delinquency.

PUBLICATION: *Crime and Delinquency*, quarterly.

Prison Families Anonymous
353 Fulton Avenue
Hempstead, NY 11550
(516) 538-6065
Executive Director: Sharon Brand

Encourages and assists families of those involved in criminal or juvenile justice processes.

Index